PROFESSIONAL
COMMUNICATION

To Merlin and Tom

PROFESSIONAL COMMUNICATION

The Social Perspective

edited by
Nancy Roundy Blyler
Charlotte Thralls

SAGE Publications
International Educational and Professional Publisher
Newbury Park   London   New Delhi

For information address:

SAGE Publications, Inc.
2455 Teller Road
Newbury Park, California 91320

SAGE Publications Ltd.
6 Bonhill Street
London EC2A 4PU
United Kingdom

SAGE Publications India Pvt. Ltd.
M-32 Market
Greater Kailash I
New Delhi 110 048 India

Printed in the United States of America

Library of Congress Cataloging-in-Publication Data

Professional communication : the social perspective / [edited by]
    Nancy Roundy Blyler, Charlotte Thralls.
        p.   cm.
    Includes index.
    ISBN 0-8039-3934-5 (cl.)   —ISBN 0-8039-3935-3 (pb)
    1. Communication in science—Social aspects.   2. Communication of
technical information—Social aspects.   3. Technical writing—Social
aspects.q.   4. Report writing—Social aspects.   I. Blyler, Nancy
Roundy.   II. Thralls, Charlotte.
Q223.P76   1993
306.4'5—dc20   92-27124
                                                                      CIP

93   94   95   96   10 9 8 7 6 5 4 3 2 1

Sage Production Editor: Astrid Virding

Contents

Foreword

CHARLES BAZERMAN

In the humanities we have a fear of the social. The humanities, we believe, constitute the place where the individual learns to express the self against compulsive society. The humanities recognize the imagination and its power to transform life rather than to submit to the social givens. We see the social opposed to the creative acts we prize and want to encourage in our students. To teach students to recognize and speak to the social is, we fear, to have them aim too low, to play other people's games, to enforce necessity rather than encourage possibility, to find a job rather than create a life.

Contemporary cultural and literary studies, even while contemplating the social, too often reproduce this terror of the social by suggesting that the social leaves no residue for the individual, for novelty. We are caught in hegemonies of our times and locales or inscribed in our socially constructed semiosis. Freedom is to be found, at best, in resistance and play.

The humanities have also defined themselves in contrast to the social sciences—those other people who claim to know something about being human, but do it in ways that threaten to diminish us by giving what we think of as reductionist scientific accounts. We have created a folklore

viii PROFESSIONAL COMMUNICATION

about the manipulative, number-crunching, positivist social sciences who have forgotten what it means to be human. This folklore helps maintain the humanities' claim to a unique and enduring role in the academy, with literature and other creative arts containing a truth not found elsewhere.

The professions suffer no better reputation than the social among the humanities, for although there are perhaps no more arcane professions than the literary theorist, the rhetorical analyst, or even the writing teacher (consider how our neighbors fear our bad opinion as much as they do dentists' and psychiatrists'), we like to think we are the champions of amateurism—just plain folks—and the integrated self that exists prior to and above modern differentiated society. We claim to offer a place where we can come to know who we are and what we mean, where the general polity can find its citizen's voice and discover its life affirmations.

But this antipathy to the social misses an important point. We are not ourselves because we set ourselves apart from each other. We become ourselves as we realize ourselves in relation to each other. The social is everything we do with each other and what we become as we do it. We individuate by identifying ourselves on a social landscape, a landscape we come to know as we interact with it. We discover and create ourselves and others by what we do with each other.

In our highly literate world, we create much of our interactive landscape through words and other symbolic representation, and we formulate our presence through those symbols that would be recognized by those around us. We become ourselves by using the common symbols for our own ends, but these ends we often discover as we interact with others. For what are most of our ends, once we get past the most rudimentary biological satisfactions, except to engage in cooperative, sharing enterprises with and for each other? As we learn to step into the complex world of society enacted in language and symbols, we find what we have to do with each other.

The developed arts and sciences, the differentiated communities and tasks of society, the specialized activities of professional life, the divisions of labor and the divisions of knowledge, the distinctive social groups and living units—these provide the locations we have established on which to make our lives with each other. It is in these socially structured locations that we make our contributions to society, find our satisfactions, gain some of our richest rewards for being part of the human experiment. The hopes and projects, needs and possibilities framed within these social nexuses are what draw out our imaginations and motivate our doing. To dread the social is to dread life.

But if the social is the stage on which all is enacted, it is a complex and protean stage. It contains law and ideas and a history of techniques of mapmaking. It contains ethics and gender and government policy. It contains character and corporations and technology. It contains small groups and large groups and people sitting alone in their rooms. It contains biochemistry and astrology and hermeneutics. It contains people who affiliate with race, ethnicity, nation, religion, and universalist ideals. It contains prisons and guns and taxes and perjury and liability. If we are to understand writing as a social act, enacted in complex social spaces, we must understand all of this.

We can no longer view writing as a limited textual practice, understood only as the bounded rules of the page. Nor is writing to be understood only as the product of an isolated mind, churning out text in like manner in any conditions as a donut machine would as long as it is in working order, in satisfactory condition and provided with raw materials. Writing is potentially responsive to and dependent on everything that is on the social stage, everything we have put there through our complex history of civilization and everything we may not have made but that we have recognized and named as being there and thus have brought into our life activity—trees and insects and electrons.

More particularly, everything that bears on the professions bears on professional writing. Indeed, within the professions, writing draws on all the professional resources, wends its way among the many constraints, structures, and dynamics that define the professional realm and instantiates professional work. The more we understand about all aspects of the professions and their situation, the more we understand about the writing that creates intersubjective places of agreement, cooperation and confrontation in the middle of all these elements.

Thus adopting a social perspective on writing, far from being reductive, requires research to gather a museum of the entire human life world. We see some of that heterogeneity in the research and theory gathered in this volume. Yet we need to understand more, to be able to recognize all the available heterogeneous elements likely to influence any particular professional writing space, to be able to see the processes by which new heterogeneous elements are borne out of evolving human practices, to understand how heterogeneous elements are yoked together into an apparently homogeneous textual meeting place, a single representation that sits in the middle of complex realms of practice. We need to recognize all that goes into local situations and opportunities and all that goes into the enduring practices of culture.

Yet while examining the proliferating human creativity and complexity that makes the most mundane of documents truly a wonder of civilization, we need also to look for fundamental human processes that might orient us to wide ranges of literate social practices. The formation of social trust and bonds; the division of interests and the creation of distrust; the processes of task definition and social enlistment; the locating of mutually recognized points at issue at socially defined times; the formation of roles, responsibilities, personalities, and tasks within complex social organizations; the creation of reasonably common understanding through processes of social typification; the use of symbols to create group cohesion or mutually perceived concerns; the means of socializing into new literate practices and the consequences of different socialization paths; the needs of individuals to have their representations accepted as worthy by reference groups; and the complications of shame (with attendant skewing of representational behavior) when one feels one's words are not appreciated—such processes of social symbolic behavior will provide a framework of understanding that will help us move from one complex communicative site to another without being overwhelmed by the infinite variety of social circumstances. Certain terms from rhetoric, literary theory, sociology, and psychology provide guideposts for our investigation of these processes: genre, *kairos,* stasis, social bond, anomie, intertextuality, socialization, typification, hermeneutics, schema, and representation. These (and many others unmentioned) suggest processes that we need to investigate further as they apply to literate behavior.

Yet no matter how many fundamental human processes we are able to identify with some specificity and regularity, these processes will never fully define what it is to write or give a comprehensive account of what one needs to do in any writing circumstance. They reveal only the social processes out of which we create the complexities and particularities of our social life. It is in these particularities that we live and to these particularities that we write.

Preface

The idea for this collection grew out of our work as co-editors of the *Journal of Business and Technical Communication* from 1985 to 1989. During that time, we witnessed a proliferation of socially based research, reflected in authors' attention to the social and cultural contexts for classroom and workplace writing and their use of quantitative and qualitative research methods for capturing the complexity of these contexts. What we were observing was, of course, not an isolated phenomenon. Other journals and collections were also reflecting and, in some cases, anticipating this shift toward what Faigley (1985) has called the "social perspective." Anderson, Brockmann, and Miller, for example, in their 1983 collection, *New Essays in Technical and Scientific Communication,* indicated that social and contextual studies will "become increasingly prominent in future research" (p. 13) on technical and scientific communication. Odell and Goswami in their 1985 *Writing in Nonacademic Settings* showed how this research might take shape through studies of writing and reading as social processes in businesses and organizations.

Working from this broad claim—that writing is a social process or act—recent scholars have further developed a social perspective on professional communication, exploring various dimensions and formulations of the social. An assessment of this scholarship suggests that in a relatively short period of time the social perspective has had a significant impact on professional communication. Perhaps most remarkable is the diversity of research that has been generated under a social orientation, as scholars have focused on a wide range of settings, relationships, and patterns concerning both classroom and workplace writing.

The collection we have put together here is designed to indicate this diversity and, at the same time, to advance our conversations about a social paradigm in professional writing research and pedagogy. These purposes are reflected in the organization of the collection, which includes two types of essays. One type is exemplified in the overview essays, which provide broad frameworks for situating developments in professional communication research as well as in pedagogy and practice. In the overviews that head part I, for example, Thralls and Blyler, and Hertzberg place current socially based research within historical and theoretical contexts, suggesting that although social views of discourse may have existed for some time they nonetheless have undergone and are still undergoing revision. Thus in chapter 1, Thralls and Blyler survey recent research in professional communication, describing three different, and to some extent competing, theoretical approaches that reveal important disagreements among researchers claiming a social orientation for their work. Hertzberg, meanwhile, in chapter 2, takes an historical perspective on social theory, tracing current social theorists' concern with local truths back to Greek Sophists and discussing more recent manifestations of the same concern in Bacon, Vico, Campbell, and Neitzche. In the overviews that introduce part II, Rymer and Freed, in separate essays, describe some impacts of a social perspective on instruction and workplace practice. Rymer, for example, in chapter 10, discusses the pedagogic impact of a social approach that assumes "discourse and education to be inherently collaborative," while Freed, in chapter 11, outlines some socially based conceptions of knowledge and their implications for organizations.

Following these overviews is a second type of essay in each section —interpretations—designed to examine and illustrate in more particular ways topics, issues, or problems that the overviews may only suggest. These essays further reflect diversity among researchers, as they interpret and use a social perspective to re-see traditional rhetorical concepts within a social context, to extend the profession's dialogue on issues important in a social perspective, and to challenge previously accepted ideas and habits in, at times, radical ways. For example, Comprone, in chapter 5, places the traditional rhetorical concept of genre in a social context, asking us to re-see genres as forms integrating both social conventions and individual intentions. Thus Comprone urges that technical writing instructors take a less mechanistic approach to the teaching of generic documents.

Berkenkotter and Huckin, Porter, Burnett, Morgan, and Kostelnick then extend the profession's dialogue on a variety of socially based issues. In chapter 6, for example, Berkenkotter and Huckin advance work on the rhetoric of science by suggesting that the need for an intertextual framework for scientists' "local knowledge" is tied to ideological and epistemological assumptions embedded in scientists' discourse practices. Porter, in chapter 7, urges that the profession's notions of ethics be expanded, arguing that ethical considerations are not matters of individual decision making but instead arise from a social matrix that involves questions of law and policy. Burnett and Morgan then focus on collaboration. In chapter 8, Burnett examines the role of substantive conflict in deferring the consensus of collaborative groups, suggesting an important connection between conflict and the effectiveness of written documents, while, in chapter 13, Morgan looks at collaborative writing tasks, offering a typology for effective assignment design. Finally, in chapter 14, Kostelnick addresses the social dimensions of visual discourse, reenvisioning the way visuals have been considered and claiming that "professional communicators must adapt pictures, just as they do text, to the experience and expectations of readers."

By challenging previously accepted ways of thinking and talking about professional communication discourse, Barton and Barton, Lauer and Sullivan, Lay, and Kent break with established tenets in ways that have the potential to reconfigure socially based research, teaching, and practice. For example, Barton and Barton, in chapter 3, and Lauer and Sullivan, in chapter 9, factor ideology into discussions of mapping practices and empirical research design, pointing to biases in the conventional ways these are viewed. Barton and Barton as well as Lauer and Sullivan thus call for researchers and teachers in professional communication to rethink established ideas about visual communication and research constructs. Similarly, in chapter 12, Lay points out the gender bias in the certain pedagogic practices in professional communication, calling for a break from traditional educational theory that reproduces this bias. And finally, Kent, in chapter 4, offers the most radical challenge to accepted ideas by bringing paralogic hermeneutic theory to bear on professional communication studies. Kent articulates a view of writing as an uncodifiable hermeneutic endeavor rather than as an act bound by communities and conventions—a view that, as Kent points out, would have profound implications for research and instruction in professional communication.

In bringing these 14 essays together, we hope this collection will generate further examination of the social perspective as a theoretical orientation for professional communication studies. In addition, by including essays that represent different, and even conflicting, ideas about the aims and purpose of research, pedagogy, and practice, we hope this collection may help point out important debates about the future direction of work in professional communication.

We are, of course, deeply indebted to all those who assisted us in preparing this collection. Our deepest appreciation goes to the authors who contributed their work and thus made this collection possible. We also thank those scholars who served as readers of individual essays at some stage in their evolution. These include Charles Bazerman at Georgia Institute of Technology, John Beard at Wayne State University, Rebecca E. Burnett at Iowa State University, Barbara Couture at Wayne State University, Linda Flower at Carnegie Mellon University, Elizabeth Flynn at Michigan Technological University, Janis Forman at the University of California at Los Angeles, Alan Gross at the University of Minnesota, John R. Hayes at Carnegie Mellon University, Lorraine Higgins at Carnegie Mellon University, Dorothy Howell at the University of North Carolina at Charlotte, Douglas Kavorkian at Meritz Communications Company, Thomas Kent at Iowa State University, Kathy Lampert at Wayland High School in Massachusetts, Juliet Langman at the University of Delaware, Janice Lauer at Purdue University, Wayne Peck at the Community Literacy Center in Pittsburgh, Berry Pegg at Michigan Technological University, Thomas E. Porter at the University of Texas at Arlington, Donald Rubin at the University of Georgia, Jerry Savage at Michigan Technological University, Brenda Sims at the University of North Texas, Patricia Sullivan at Purdue University, and David Wallace at Iowa State University.

We are also grateful for the assistance of Ann West, our acquisitions editor at Sage Publications, and for the comments of Carolyn R. Miller and Jack Selzer, who reviewed the prospectus for this collection.

Finally, we wish to acknowledge those in the Department of English at Iowa State University who helped us complete this project. We especially want to thank our colleagues in the Rhetoric and Professional Communication Program at Iowa State. By offering us their encouragement and guidance, this group helped create for us an environment supportive of the editing and research that went into this project. Special thanks also go to Sheryl Kamps and Norma Michalski, who provided invaluable clerical assistance from the proposal through the final manuscript stages of this project, to Ray Brautch, our graduate research assistant who helped prepare the bibliography for this collection, and to the department for its financial support of our research.

—*Nancy Blyler*
Charlotte Thralls

PART I

History, Theory, and Research

OVERVIEW

1

The Social Perspective and Professional Communication: Diversity and Directions in Research

CHARLOTTE THRALLS
NANCY ROUNDY BLYLER

The 1980s have seen a growing interest in socially based research, for both composition (Nystrand, 1989, p. 66) and professional communication. In conjunction with this increased interest, writing theorists have attempted to articulate the shared assumptions that underlie what Faigley (1985) termed the *social perspective*. Perhaps most binding is the fundamental rejection of positivism and the windowpane theory of language. According to positivism, knowledge is a direct apprehension of reality, where the human mind acts as a mirror reflecting the outer world, and ideas are true to the degree that they correspond with the world. The function of discourse, therefore, is to transcribe, as clearly and accurately as possible, what the mind as windowpane has apprehended. (For a fuller discussion of positivism, see C. R. Miller, 1979.) Social theorists reject these positivistic notions, arguing instead that reality is unknowable apart from language.

Because social theorists further hold that language and culture are intimately related, they share a belief in the importance of the communal. Faigley, for example, claimed that research employing a social perspective examines "how individual acts of communication define, organize, and maintain social groups" (1985, p. 235) and that writing "can be understood only from the perspective of a society rather than a single individual" (1986, p. 535). Echoing Faigley's emphasis on social groups, Carter (1990, p. 266) stressed that socially based research focuses on the *local*—a term borrowed from Geertz (1983) for the dependence of knowledge and writing on individual communities. Although Nystrand (1989) took social theorists to task for imprecision about the ways communities inform writing, he too shares Faigley's and Carter's emphasis on social groups (pp. 66-73).

This rejection of positivism, in conjunction with social theorists' stress on the local and communal, has led to one additional claim: the centrality of socially mediated meaning. Because there is no immediate knowledge of reality and because both knowledge and discourse are bound up with specific social groups, communications are invested with meaning only through the interactions of writers and readers in those groups. In short, socially mediated meaning—or, to use an alternate term, *interpretation*—is central to the social perspective.

These common characterizations of the social perspective—as a departure from positivism and as a theory concerned with the local, the communal, and the social mediation of meaning—connect a social perspective on writing research with social views of discourse prevalent in a range of current theoretical movements: poststructuralism, radical feminism, and the philosophy and sociology of science. This widespread theoretical turn toward a social view of discourse has had a particularly strong impact on our understanding of the human sciences. For example, the authors in two volumes—*The Rhetoric of the Human Sciences* (Nelson, Megill, & McCloskey, 1987) and *Rhetoric in the Human Sciences* (Simons, 1989)—characterized knowledge and disciplines in terms of the communal and local, seeing field-dependent discourses as central to the definition of the various human sciences.

It appears, then, that some shared presuppositions are evolving to characterize a social perspective on writing research and that this research is a part of a larger cross-disciplinary movement to contextualize discourse within social and cultural influences. It appears as well that a primary focus among rhetoricians and writing researchers has been reiterating and elaborating on the common tenets of this perspective. This emphasis on commonality now presents us with two problems. The first has to do with the degree to which the social perspective can be characterized as a monolithic paradigm. When Faigley outlined this perspective, he indicated various directions that socially oriented research might take, but his chief concern seemed to be differentiating the social from other more entrenched perspectives—the textual and the individual (1985) or the expressive and cognitive (1986). As a result, Faigley emphasized the social perspective as a framework for uniting socially based writing research under a single theoretical orientation. Now that the social perspective has taken hold, it has become increasingly apparent that significant differences exist among socially oriented theorists and researchers. How then do we account for these different assumptions and emphases that are beginning to emerge among rhetoricians and writing researchers, all of whom claim a social foundation for their work? Stated another way, how can we begin to describe the different approaches *within* the social perspective?

A second and related problem has to do specifically with professional communication research. Thus far, we have no overview studies describing the diversity of work in this area being done under a social rubric—including important theoretical variations within a social perspective. Because these variations in social theory are now clearly beginning to emerge in studies of workplace writing, how can we map this rapidly developing body of socially based research? How can we make sense of existing research and assess the directions that socially based studies in professional writing seem to be taking?

Given these two problems, the purposes of this chapter are twofold. First, we want to describe three different and, to some degree, competing theoretical approaches that have been developing within a social perspective: the social constructionist, the ideologic, and the paralogic hermeneutic. Of course, we do not claim that these three approaches represent all the possible ways socially based research might be discussed or exhaust all the possible differences in socially based writing research. We do not, for example, account for research with alliances to other theoretical orientations, such as the cognitivist and the expressivist, or for research that embeds rather than makes explicit its theoretical presuppositions. We do, however, believe that the three approaches we discuss allow us to identify some competing assumptions about writing as a social act, thus permitting us to get at crucial issues that divide researchers operating within a social perspective. These differences are important, we believe, because differing assumptions about the *social* are leading to distinct emphases and bodies of research in professional communication.

Our second purpose is to use the social constructionist, the ideologic, and the paralogic hermeneutic approaches as ways to conceptualize some important patterns and differences that are emerging in studies of professional writing. As our discussion suggests, a social constructionist approach has dominated professional communication research, but the ideologic and paralogic hermeneutic approaches are beginning to offer important challenges to social construction. By exploring the implications that all three approaches have for professional communication research, we hope to characterize dominant patterns in existing research and point out possible directions in future research.

In the following sections we explore each of the three approaches that we see emerging within the social perspective, using four concepts: *community, knowledge and consensus, discourse conventions,* and *collaboration.* We have chosen to concentrate on these concepts because they are key terms framing discussion and debate among social theorists. To illustrate different ways these concepts are understood and employed under each approach, we cite representative studies in professional communication, and in the case of the ideologic and paralogic hermeneutic approaches, we show how these two approaches are generating critiques of research that has been informed by social constructionist principles. In the final section, we assess the major contributions of each approach, and we speculate on the challenges offered to professional communication research in light of ideologic and paralogic hermeneutic theory.

The Social Constructionist Approach

Social construction is perhaps the best known and certainly the longest reigning of the three approaches that comprise the social perspective. Bruffee's (1986) seminal article, "Social Construction, Language, and the Authority of Knowledge: A Bibliographic Essay," traces his social constructionist approach to such theorists as Kuhn (1970), who claims that scientific knowledge is "the common property of a group or else nothing at all" (p. 201); Rorty (1979), who rejects knowledge as representation, endorsing instead the view that knowledge is "the social justification of belief" (p. 170); and Geertz (1983), who views all knowledge as local (p. 4). In doing so, Bruffee consolidates the work of these and other theorists, summarizing their major points and discussing the implications of social construction for education in the liberal arts.[1]

The cornerstone of Bruffee's (1986) constructionist approach is the belief that knowledge is not "individual, internal, and mental" but instead is social in nature (p. 775). In endorsing this tenet, Bruffee sees himself breaking with a long-standing cognitivist tradition, rejecting epistemological assumptions that posit a universal foundation for knowledge. Instead, Bruffee (1986) and the constructionists he cites claim that knowledge is nonfoundational, emerging from a social matrix (pp. 776-777). This matrix is defined, in Bruffee's social constructionist approach, by the concept of community.

Community

Community is central to social construction because communal entities are the sources of knowledge. As Bruffee (1986) asserts: "Social construction understands reality, knowledge, thought, facts, texts, selves, and so on as community-generated and community-maintained linguistic entities" (p. 774).

Despite the importance of community in this approach, social constructionists have yet to agree on precisely what the term means. Although Bruffee (1986, p. 784) aligns it with Fish's (1980) notion of an interpretive community, other theorists point to variations in the constructionist understanding of community. One variation, which stems from sociolinguists' concept of a speech community, designates a more specific group, whose members are often circumscribed by physical location (J. Harris, 1989, p. 14) and who share norms of behavior and communication (I. Thompson, 1991, p. 42). A second variation popular in composition and in professional communication—that of the discourse community—usually retains some of the specificity inherent in the sociolinguistic concept of a speech community (J. Harris, 1989, pp. 14-15) and describes a group whose members "acquire specialized kinds of discourse competence that enable them to participate" in the community (Faigley, 1985, p. 238).

Despite these variations on the precise meaning of community, theorists point to a common thread: the presupposition of like-mindedness on the part of community members (J. Harris, 1989; Kent, 1991). Discussed by Bruffee (1986, p. 777) as a characteristic of communities of knowledgeable peers, this like-mindedness takes concrete form in shared communal beliefs, values, and practices (J. Harris, 1989, p. 15). To social constructionists, these shared beliefs then manifest themselves as norms or standard, consensually held assumptions of community members that shape the discourse the community produces (Rafoth, 1988, p. 140). In addition, however, to communities' shaping discourse, Bruffee (1986, p. 774) also claims that the community is, in turn, shaped by the discourse it has generated.

This constructionist concept of community has widely influenced scholarship in professional communication, where the concept of a discourse community has given researchers a way to talk about workplace writing in both industrial and academic settings. In particular, researchers have examined the normative aspect of communities, focusing on the way these regulate discursive practices. Concerning industrial communities, for example, Freed and Broadhead (1987) study a management consulting firm and an international accounting firm as discourse communities whose "institutional norms" "reign over" or "legislate" the writing done in those settings (pp. 156-157). Odell (1985) also suggests that the state bureaucracy he studied acted as a normative community, wherein shared attitudes, previous actions, and ways the organization routinely functioned were used to justify members' writing choices (p. 252). Similarly, Driskill (1989) speaks of corporate culture—values, norms, and beliefs that guide action—as the source of the "interpretive standards that affect writers' choice of content, persuasive approach, and word choice" (p. 137).

Concerning academic settings, Myers (1985) examines the proposals written by two biologists, concluding that both proposals were shaped significantly by the expectations of reviewers, expectations that caused the proposal writers to fit themselves within the discipline's consensus about acceptable research (p. 237). In their study of student writers, Anson and Forsberg (1990) find that students making the transition from the classroom to the workplace must also develop "strategies for social and intellectual adaptation" to the contexts provided by different discourse communities (p. 202).

In addition to examining the normative influence of communities, researchers in professional communication have studied the way industrial and academic communities are in turn shaped by their discursive practices and the discourses they generate. Doheny-Farina (1986), for example, examines the collaborative writing of a company's business plan, finding that "the writing process helped to resolve the power struggle, shaping the social reality of the top management of the company" (p. 178) and altering the way the company viewed its authority structures. Similarly, Journet (1990) studies the case histories written by two neuro-psychiatrists, concluding that by incorporating narration into the more traditional, analytic modes of neurological writing these professionals expanded the definition of what it meant to be a neuropsychiatrist (pp. 179, 182-183, 194).

Thus the social constructionist approach has made community an important focal point in professional communication research. In particular, community is both a normative force, the origin of the shared values and beliefs that shape and even regulate communication, and subject to being shaped by the communicative process and discourse the community generates. The shared values and beliefs of community members then take on additional significance because these values and beliefs define the community's consensus about knowledge.

Knowledge and Consensus

In Bruffee's social construction, knowledge results from a community's consensus about what it will call true, rather than from a universal that will ensure truth. As Bruffee (1986) asserts: "There is no such thing as a universal foundation, ground, framework, or structure of knowledge. There is only an agreement, a consensus arrived at for the time being by communities of knowledgeable peers" (pp. 776-777). Following Rorty, Bruffee (1986) then terms such consensual knowledge socially justified belief (p. 780).

This concept of knowledge as a consensually held, socially justified belief has influenced research in professional communication by shifting attention away from universals that ensure truth and toward the means by which such beliefs are incorporated into a community's knowledge store. One such means is interaction among community members, which researchers suggest can lead to widely shared agreement. Spilka (1990), for example, studies engineers in a large corporation, concluding that orality—"the process of transmitting ideas via any conversation or message between project participants that involves speech" (p. 44)—was a central means of achieving "a corporate consensus" concerning, among other things, the contents of documents (p. 45). Similarly, in examining the *Challenger* tragedy, Winsor (1990a) points to the impact that more effective communication among engineers and managers might have had on incorporating ideas about the faulty O-rings into the managers' knowledge store (pp. 17-18).

Winsor (1990a) asserts that interaction enables consensus by offering the opportunity for community members to come to mutual understanding (p. 18). At the same time, however, interaction may have a regulatory effect, limiting what will be admitted into a community's body of knowledge. For example, in two studies of engineers, Winsor examines the role of inscription—a term borrowed from Latour and Woolgar (1986) to describe the "encoding of experience in socially validated symbols" (Winsor, 1989, p. 271)—in the social construction of knowledge. Concluding that "knowledge equals text," Winsor (1989, p. 284) points to the way in which the texts engineers consult and write, act as inter-text, constraining the knowledge generated. Similarly, Berkenkotter and Huckin (this volume) posit that a framework of accepted knowledge, as revealed through citation, provides essential intertextual support for a claim to scientific discovery. Without this support, Berkenkotter and Huckin suggest, such a claim will not be admitted as knowledge. And finally, in a study of physicists' reading habits, Bazerman (1985) discovers that the physicists' interactions with other research and their judgments of it influence and constrain "the course of the whole community's knowledge" (p. 15).

This focus on knowledge as consensual agreement and the means by which beliefs are incorporated into a community's knowledge store has enriched the constructionist understanding of the functioning of communities by indicating how knowledge might be maintained and, to some extent, how it might grow. Concerning this growth, however, researchers in professional communication have paid more attention to the slow accretion of knowledge through interaction than they have to radical changes or shifts in a community's beliefs. Following Rorty, Bruffee (1984) links such changes to *abnormal discourse,* or discourse challenging the prevailing community consensus about knowledge (p. 648), but few studies in professional communication have followed Bruffee's lead in examining such challenges. Instead, researchers in professional communication have focused on knowledge in relation to Bruffee's (1984) *normal discourse,* or the discourse that members of communities most often write (p. 642). In constructionist theory, this normal discourse can be identified by the conventions the community endorses.

Discourse Conventions

Discourse conventions play a central role in social construction because, according to Bruffee (1986), communities are constituted by the language their members employ (p. 779). Thus communities are defined and set off from one another by their discursive practices, which allow them to justify their beliefs socially and arrive at consensus about what they will call knowledge (Bruffee 1986, pp. 778-779).

Because communities are constituted by language and because beliefs are justified socially through language, the discourse conventions characterizing communities have received much attention in professional communication research. Perhaps the simplest form this research has taken is the study of the conventions that identify various communities. Herrington (1985), for example, examines two chemical engineering classes, determining that these "forums" were "constituted by distinct intellectual and social conventions" (p. 405), which students could then learn. Similarly, in studying the discipline of engineering, C. R. Miller and Selzer (1985) note that engineering discourse uses classical *topoi* or topics specific to genres, to organizations, and to the discipline as a whole, which are related to "community-specific conditions of successful argument" (p. 311). Finally, by examining the introductions of articles written for several different interpretive communities within psychology, Walzer (1985) concludes that "a writer's audience should be thought of in terms of the conventions of the discourse" of those communities (p. 157).

On a more complex level, however, professional communication research has been concerned with discourse conventions as indices of community membership. In particular, researchers have noted the relationship of conventions to communal norms. In a study of auditors' writing, for example, Hagge and Kostelnick (1989) determine that negative politeness strategies were a response to the norms of the auditors' firm. Similarly, Brown and Herndl (1986) find that professionals in 15 corporations used nominalizations and narrative structures because "these language features had acquired . . . powerful and favorable significance as signs of group affiliation" (p. 13). Researchers, however, have also noted the regulatory effect discourse conventions can have, delineating the parameters of the community and constraining its members. To Lipson (1988), for example, internalized conventions of discourse reinforce the values of a profession and thus serve to perpetuate it. "Through language specialization," Lipson (1988) claims, "the professional internalizes a control of behavior that reminds the professional of the value of adhering to the special norms of the specialized group" (p. 9).

This constructionist view of discourse conventions places the utmost importance on language as the means by which communities are constituted. Because discourse conventions are thus intimately tied to communities and community membership, Bruffee and other constructionists have stressed ways nonmembers can internalize both community norms and language and acquire membership. Bruffee has focused specifically on collaboration as a means to facilitate these goals.

Collaboration

Social construction posits collaboration as both the social process implicit in all writing and a pedagogic tool for teaching writing. As a social process, collaboration refers to what Bruffee (1984) calls "the conversation of mankind," whereby "thought is internalized public and social talk . . . and writing of all kinds is internalized social talk made public and social again" (p. 641). As "internalized conversation re-externalized" (Bruffee, 1984, p. 641), writing thus is not a solitary but a communal and collaborative act. Writing is learning to participate in conversations by fitting one's talk to the demands of knowledgeable peers, those "who accept, and whose work is guided by, the same paradigms and the same codes of values and assumptions" (Bruffee, 1984, p. 642).

For Bruffee and other social constructionists, collaborative learning —peer critiquing and tutoring, reader-response groups, and group-writing projects—is a tool for students to learn how to participate in the conversations of these knowledgeable peers. Organizing students into peer groups, Bruffee maintains, can model the normal working of discourse communities and the social nature of authorized knowledge. Giving students collaborative tasks that require them to use the normal discourse of groups can further enable students to learn the conversational values that will effect their transition into professional communities. Bruffee also expresses interest in collaborative learning and writing as methods for de-centering authority in the classroom and for promoting democratic decision making, but his primary concern with collaborative pedagogy is its acculturative value.

Although Bruffee's vision of collaborative learning is only one of many influences on collaborative pedagogy as it has emerged during

the 1980s, Bruffee's emphasis on collaboration as facilitating acculturation has had a significant impact on collaborative work in professional
communication, providing an important rationale for collaborative instruction and research as well as grounds for a cooperative relationship
between academe and industry.

In terms of classroom collaborations, many researchers have examined how collaborative writing projects can help acculturate students to
their academic disciplines and professions. In their early work on
collaboration, for example, Forman and Katsky (1986) argue that collaborative writing projects in management communication courses teach
students how to talk within "frameworks" of their disciplines, thus
enabling them to become effective members of marketing, finance, and
business economics groups (p. 32). In a more recent study, Rogers and
Horton (1992) examine the importance of face-to-face collaborative
interaction in order to foster the "'talk about talk' that Bruffee deems
necessary for individuals to learn the discourse of a community" (p. 122).

The idea that collaboration can facilitate students' transitions to the
communication demands of professional discourse groups has lead to
an even greater body of research on collaboration in the workplace.
Arguing that collaborative writing is a norm in business and industry,
many researchers (e.g., Farkas, 1991; Selzer, 1989; Van Pelt & Gillam,
1991) have studied collaboration in nonacademic settings to determine
the types of collaborative arrangements and strategies employed in the
workplace, with an eye toward incorporating this information into the
classroom. The underlying aim of this research thus is largely acculturative: Studies of workplace collaboration yield knowledge about professional practice that can then be used to prepare students to function
effectively in their jobs.

Because collaboration viewed as an acculturative activity has been
used by many teachers and researchers to link the classroom with the
workplace, a constructionist approach to collaboration has helped establish—or, at minimum, reinforce—the possibilities of a cooperative
relationship between academe and industry. Seen as an area of mutual
interest, collaboration in the classroom and collaboration in the workplace become complementary activities, with classroom collaborations
fostering interpersonal skills that will help students make effective
transitions to the workplace.

To summarize, the social constructionist approach focuses on community, viewing communal entities as the sources of knowledge maintained by consensual agreement; as the repositories of discourse

conventions by which communities are defined and shaped; and as the bodies to which nonmembers must—through collaboration—be acculturated. The ideologic approach focuses on political issues downplayed in constructionists' ways of conceptualizing community, knowledge and consensus, discourse conventions, and collaboration.

The Ideologic Approach

For those reacting to and critiquing social construction, social theory has generated renewed interest in rhetoric as ideology. The emphasis here must be on *renewed* because attention to rhetoric as a political act is, of course, not a new concern in writing research. Expressivists (i.e., Elbow, Murray, Coles, Macrorie), have been, for example, deeply interested in the rhetoric of power and the responsibilities of individuals to challenge established institutional order. More recent scholars, such as Berlin and Myers, share the expressivists' interest in issues of power but reject as romantic expressivists' tendencies to view individuals as separate or free from the social and institutional forces that shape individual identities. Current scholars interested in the ideological dimensions of rhetoric see writers and the aims and intentions of writing as socially structured by ideology because, in Berlin's (1988) words, "notions of the observing self, the communities in which the self functions, and the very structures of the material world are social constructions" (p. 488).

Because of this emphasis on community and the role of discourse in constructing reality, current scholarship that takes an ideologic approach should be understood as an extension and elaboration of, rather than a major departure from, Bruffee's constructionist theory. Scholars interested in the ideologic dimensions of discourse see themselves factoring ideology into the tenets of social construction as a way of correcting crucial oversights in constructionist theory. More specifically, these scholars believe that Bruffee and his followers in writing research have separated the social conditions of writing from questions of power and control, and thus constructionists have tended to characterize knowledge making as a benign and apolitical process. Scholars such as Myers, Berlin, and Trimbur are concerned that when ethnographic studies and empirical surveys merely describe how writing practices shape and are shaped by social forces within academic and business communities, these studies represent community standards and values as normative, already-in-place, even unauthored givens to which new members must be assimilated. Such studies thus risk misrepresenting social arrangements as normal and natural, rather than as practices kept in place by power structures within institutions.

As a corrective, scholars taking an ideologic approach wish to extend social inquiry to include the ideologic frameworks that shape language practices and thus thought and identities within professional communities. By pointing out the systems of power and the means of production that authorize knowledge within discourse groups, these scholars wish to demystify the structure of authority behind knowledge-making processes. They also wish to advance a liberationist agenda: to empower members of communities to shape the discourse of their groups rather than to be passively shaped by the discourses of those controlling the production and distribution of knowledge.

Because this ideologic approach calls into question some of the research and pedagogic practices in professional communication as well as the values of the commodity culture underlying many business and scientific communities, only a handful of scholars identified with professional communication are looking directly at rhetoric and professional communication practices through an ideologic lens. For the most part, ideology-based scholarship is coming from rhetorical theorists such as Berlin, Myers, Miller, Bizzell, and Trimbur, all of whom, to some degree, have been influenced by postmodern thought and leftist literary theory. Together, these scholars, by critiquing and extending constructionist concepts of community, knowledge and consensus, discourse conventions, and collaboration, are opening new avenues for research in professional communication.

Community

An ideologic approach is concerned with the authority structures that enable communities in business, industry, and the professions to maintain and legitimize social orders and practices under the auspice of tradition and custom. Leftist critics refer to this self-legitimizing function of communities as *reproductive ideology*—a term derived from Marxist theory to describe the process whereby communities control and perpetuate themselves. According to Myers (1986), whose work since 1986 has taken an overtly ideologic turn, academic and corporate communities are powerful mechanisms of reproductive ideology because structures within communities tend to distribute power in ways that protect practices supportive of empowered social and economic groups. In protecting these structures, communities create dominant ideologies which, Trimbur (1989) argues, normalize "hierarchical relations of power (p. 603)," reducing "the authority of knowledge to a self-legitimizing account" (p. 609) of community practices. Academic and professional communities thus create, Trimbur concludes, "monopolies of knowledge" that privilege and reproduce "the meritocratic order of credentialed society" (p. 611).

Communities as products and reproducers of ideology have been the focus of considerable research in general education and composition. Bartholomae (1985), Friere (1968/1983), Giroux (1983), and Knoblauch (1988), for example, show how traditional reading and writing instruction is an entrenched pedagogic practice in academic communities, reproducing and perpetuating patterns of authority and social class. In contrast, research in professional communication is only beginning to examine the ideologies of corporate cultures and the role that pedagogy and research can unwittingly play in reproducing the values and ethics of dominant groups.

Feminist critics in professional communication and organizational culture, for example, are scrutinizing the way in which academic and business communities reproduce sexism and hierarchical social arrangements. Acker (1991), in her study of organizational culture, shows how a gendered substructure is reproduced through the written work rules, labor contracts, managerial directives, and other documents communities use to describe and evaluate jobs. Lay (this volume) examines how these gendered and hierarchical values are reproduced in professional communication classrooms through reproductive educational theory that reinforces gender distinctions and male dominance.

C. R. Miller (1989) issues perhaps the strongest critique of professional communication pedagogy and research as sources of reproductive ideology, perpetuating the ethics of commodity culture. Miller, for example, questions the ideology that is reproduced through emerging "industry-university collaboration" (p. 19) and through research on work-related writing. Uncritically importing into the classroom the communication processes and practices of industry reproduces private corporate interests, making students the tools of capitalist ideology. This reproductive process, Miller argues, allows corporate ethical and political values to be absorbed and passed on without students reflecting on the impact that private communities may have on the good of the larger public community.

As these studies suggest, a leftist critique of *community* is leading researchers to a more self-conscious examination of the larger cultural and economic contexts in which professional communication research and pedagogy take place. An ideologic critique also, however, calls for careful scrutiny of the internal workings of communities and the production apparatus that enables communities to control and perpetuate themselves. For those taking an ideologic approach, this production apparatus is consensus, which supports the prevailing knowledge.

Knowledge and Consensus

In the ideologic approach, consensus is an instrument of power and exclusion. In associating consensus with exclusion, ideologic critics depart from Bruffee's analysis of consensus—what Myers (1986) calls "the weak point" (p. 166) of his theory. At issue is Bruffee's willingness to concede almost complete authority to the accepted knowledge of community members while minimizing the discourses that are silenced or excluded in the consensus-making process. For Bruffee (1986), conflicting views and individual differences can be explained either as views that must be eliminated or as views that can be reconciled through rational negotiation and thus be absorbed into a prevailing consensus (pp. 647-649).

For those adopting an ideologic approach, consensus so narrowly construed masks community conflict and heterogeneity—egocentrism and gender, individual and class differences—and downplays the fact that "knowledge and its means of production are distributed in an unequal, exclusionary social order and embedded in hierarchical relations of power" (Trimbur, 1989, p. 603). In short, some interests are suppressed while others dominate. Consensus thus is not so much an index of agreement as an exercise of power, resulting in what Habermas (1970, p. 205) calls a "systematically distorted communication" (a false consensus) and Bachrach and Baratz (1962) term a "mobilization of bias" (p. 950).

Influenced by such analyses of consensus, researchers in rhetoric and writing and in organizational culture are now attempting to demystify consensus by situating it within the larger context of conflict, contradiction, and difference. Myers (1986) argues, for example, that "we need to see consensus . . . as the result of conflicts, not as a monolith"

(p. 166); while Mumby (1988) argues for a perspective that will examine "the underlying structural conditions that produce situations of consensus or coercion (or coercively produced consensus)" (p. 48). Trimbur (1989) has labeled this perspective a consensus of "opposition" or "dissensus" (pp. 609-612). As a critical strategy, dissensus foregrounds voices that are left out or suppressed in the control and distribution of knowledge. A rhetoric of dissensus, Trimbur (1989) explains, would "open gaps in the conversation through which differences may emerge" (p. 614) and identify the "forces which determine who may speak and what may be said" (p. 612).

An ideologic critique of consensus has important implications for research and pedagogy in professional communication. For example, a consensus of opposition challenges constructionist-oriented studies that suggest that the disciplinary practices of biologists, lawyers, engineers, and other professionals reflect consensual agreement. Consensus, J. Harris (1989) maintains, is not necessary in professional communities (p. 20) and not necessarily, Trimbur (1989) argues, the norm in business and industry (p. 610). A consensus of opposition further challenges the notion that professional documents reflect consensus when, in fact, such documents may reflect power struggles and contradictions that prohibit consensus.

Wells (1986) is among the few researchers in professional communication to address the myth of consensus in industry and the professions. Through her analysis of manual writing in an industrial setting, Wells illustrates how conflict can actually impede consensus, and she suggests how awareness of these conflicts may revise traditional conceptions of technical documents and the technical writer's role. By illustrating how these conflicting interests result in three separate documents that suppress and segment information, Wells points out the extent to which technical documents may reflect and even preserve competing claims to authority.

As Wells's study and Trimbur's concept of dissensus suggest, an ideologic approach to knowledge and consensus is directing attention away from constructionists' notions of this concept as indicating agreement among community members and toward the relations of power that authorize some knowledge claims and exclude others. In so doing, an ideologic approach also redirects the analysis of discourse conventions because, if consensus is the production apparatus for reproducing communal values, discourse conventions reflect and reify that consensus.

Discourse Conventions

An ideologic approach identifies discourse conventions as complex semiotic systems or symbolic orders that signify and sustain the relations of power implicit in consensual knowledge. Scholars adopting this approach wish to deconstruct this symbolic order, to lay bare the struggle between dominant and "socially disenfranchised discourses" (Bizzell, 1986b, p. 43)—to show how textual norms and ways of reasoning can demarcate insiders and outsiders, privilege what can be said and how, and exclude or marginalize alternate discourses.

An ideologic approach to discourse conventions has significant implications for the direction and focus of professional communication research and pedagogy. Perhaps most important, this approach shifts focus away from constructionists' preoccupation with conventions as identifying both communities and community members and toward the ideologies that underlie discourse and the ways in which conventions socially construct relations of domination. Researchers in professional communication are beginning, for example, to examine how text conventions in technical documents can be used to serve dominant ideologic interests. Thus, in their study of environmental impact statements (EISs), Killingsworth and Steffens (1989) illustrate how the mandated discourse of EIS reports, while ostensibly a "democratizing rhetoric" (p. 170), actually protects the interests of the government bureaus, thereby preserving the dominance of this group.

Other researchers are looking at even more subtle ways that seemingly neutral discourse elements can mask ideology. Barton and Barton (this volume), for example, " 'deconstruct' the innocence of maps." By outlining "rules of inclusion" and "rules of exclusion," Barton and Barton illustrate how maps privilege and reproduce the interests of dominant groups at the same time that maps naturalize the relationship between signifier and signified and thus "dissimulate" the fact of privileging. In addition, Lauer and Sullivan (this volume) argue that validity and reliability "are not transcendent, self-evident characteristics inherent in the nature of empirical research" but instead reveal studies' theoretical assumptions and hence betray ideology.

Finally, an ideologic analysis of discourse conventions is leading researchers in professional communication to reevaluate marginalized discourses and assert the legitimacy of alternate, excluded patterns of reasoning and writing. In separate studies, for example, both Brodkey (1987) and Rogers (1989) have called for a reevaluation of narrative in professional discourse. Thus Brodkey addresses the suppression and exclusion of narrative in academic research, with its predilection for data gathering and analysis, arguing for greater reliance on narrative in ethnographic and qualitative research because of the "lived experience" (p. 40) that narratives provide. Rogers makes a similar argument in her study of automobile dealer contract reports. Although official guidelines for these reports recommend against narrative as an organizational pattern, Rogers shows that narrative is preferred by managers and better fulfills evidentiary requirements of the company.

Taken together, many of these studies represent a shared point of view being advanced by an ideologic analysis of professional discourse. Killingsworth and Steffens (1989), and Brodkey (1987), for example, emphasize what Barton and Barton call "the perspective of the traditionally disempowered" in order to encourage alternative and more heterogeneous discourses. At issue for these authors is resistance to reproductive ideology and empowerment of those voices and vocabularies slighted by conventional and privileged ways of speaking and writing. This concern with resistance and empowerment also informs an ideologic approach to collaboration, marking important factors that ideologic critics find missing in social constructionist pedagogy.

Collaboration

Influenced by the educational and political theories of Freire (1983), Giroux (1983), Habermas (1970), and others, scholars in rhetoric and composition are attempting to broaden the constructionist agenda for collaborative writing and learning. Collaboration, these scholars argue, should demonstrate to students not merely that knowledge is socially constructed; collaborative activities should also, in Trimbur's (1986) words "change the social character of production" (p. 612). In other words, collaborative learning should enable students to participate in decision making and thus be empowered to control their situations.

Knoblauch (1988), for example, has advanced Freire's concept of problem posing, a critical strategy that would help students become aware of the shaping influence of interpretive communities "without subordinating themselves to entrenched ideas about their proper place or function within the institution of 'literacy' " (p. 135). Trimbur's dissensus—modeled after Habermas's (1970) ideal speech situation, a utopian ideal—would rearrange collaborators' interactions to achieve relations of nondomination. For ideologic critics, collaboration thus is a way to transform the classroom: to break down authority structures so that teachers and students collaborate with shared authority, to foreground the voices that are lost as student collaborative groups merge their collective views in writing projects, and to draw out the hegemony implicit in prevailing ideas about appropriate discourse in various disciplines and professions.

This concern with the ideologic dimensions of collaborative activity is beginning to emerge in professional communication, constituting what Forman (1992) calls an important "new vision" for collaborative writing research (p. xvi). In their studies of workplace writing, for example, researchers like Trimbur and Braun (1992) and Locker (1992) are starting to look at the distribution of power and authority among co-authors and within collaborative teams. In studies of collaborative pedagogy, meanwhile, researchers are addressing how power might be distributed more equitably among students in collaborative groups and between students and teachers in technical and business writing classrooms. In their study of engineering course collaborations, for example, Flynn, Savage, Penti, Brown, and Watke (1991), build on Ede and Lunsford's (1990) hierarchical and dialogic modes of collaborative interaction to show how these modes may be gender marked and linked to the distribution of power relations. Associating a hierarchical mode with dominance and oppression (male collaborative style) and a dialogic mode with equality and balance (female collaborative style), Flynn and co-authors argue for the more equal distribution of power that the dialogic mode provides. They also introduce a third mode of collaborative interaction—asymmetrical—designed to account for the different responsibilities of group members and to protect against devalued responsibilities, particularly those that may be assumed by women.

Other researchers are examining how new technologies might help students better understand, and possibly even dissolve, hierarchical structures in collaborative interactions. For example, Rymer (this volume) shows how videotaping a writing group during the planning stage of its collaborative work can empower students by allowing them to witness the conflicts and power struggles involved in their movement toward consensus. Many researchers are also finding great promise for egalitarian reform in the computer-supported collaborative classroom. Separate studies by Arms (1987), Selfe and Wahlstrom (1989), and Duin (1991), for example, suggest that computer-supported classrooms allow for flexible authority structures, transforming teachers into coaches and collaborators and student collaborators into empowered evaluators. In their study of computer networking (electronic bulletin boards and teleconferencing), Kiesler, Siegel, and McGuire (1988) conclude that networking fosters democratization because the anonymity of networking interchanges eliminates many cues of status and authority.

Technology in the collaborative classroom is itself, however, being subjected to an ideologic critique. Recently, for example, Hawisher and Selfe (1991) have cautioned that, although computers have the potential to "help us shift traditional authority structures" (p. 62), computer networking systems could ultimately help reproduce repressive structures by allowing the instructor to monitor and control students' collaborative interchanges.

In summary, ideologic critics have differed significantly with the social constructionist understanding of community, knowledge and consensus, conventions, and collaboration by pointing to the tendency of communities to reproduce their ideologies and thus to suppress difference and by examining the pedagogic implications of and alternatives to existing structures of authority. Paralogic hermeneutics, a third social approach and one that has only recently made itself felt in writing research, also departs from social construction. The disagreement voiced by paralogic theorists, however, is more fundamental than the disagreement voiced by those following an ideologic approach, because paralogic hermeneutics reinterprets in profound ways many of the tenets on which social construction is based.

The Paralogic Hermeneutic Approach

Paralogic hermeneutics, which has sparked interest among a small number of writing theorists who argue with current views of discourse production, derives its name from its concern with the interpretive, or hermeneutic, act. Of course, paralogic hermeneutics is not alone in this concern with interpretation, because both the social constructionist and the ideologic approaches also endorse the social mediation of meaning. Unlike these approaches, however, paralogic hermeneutics stresses the uncodifiable—and hence the paralogic—nature of interpretation. (Kent, 1986b, should be credited with linking the term *paralogic* to hermeneutics, p. 25.) This uncodifiability derives from what has been called *externalism* and a subsequent emphasis on the primacy of communicative interaction.

Paralogic theorists claim all interpretation and hence all understanding arise directly out of communicative interaction. These theorists, therefore, reject what they identify as a central presupposition of most writing theory and research: the *internalist* presupposition that a Cartesian split exists between the human mind and whatever is outside of the mind—a split that must be mediated by "some sort of epistemological network" internalized and employed to gain knowledge of reality (Kent, in press-a). In rejecting this internalist idea, paralogic theorists also reject what they see as a central constructionist tenet: the notion that, because communities are constituted by their languages, being a member of a community and internalizing the communal language are prior requirements for understanding (Kent, this volume, in press-a). Citing a long philosophical tradition, paralogic theorists assert that "no split" requiring mediation "exists between an inner and outer world" and claim instead that understanding develops out of "the give and take of communicative interaction" (Kent, in press-a). This view of the relationship between the mind and reality has been called externalism.[2]

Because paralogic theorists believe that understanding comes through communication, the interpretive or hermeneutic act can never be outlined in advance of an interaction. As Kent (1989b), based on the work of Davidson and Bakhtin, asserts, this act is open-ended and dialogic (pp. 25, 37) and, therefore, cannot be determined by the language practices of communities. For paralogic theorists, the hermeneutic act is truly uncodifiable (Kent, 1989b, p. 25).

Because paralogic theorists essentially reject tenets they see as central to social construction and its ideologic critique, paralogic hermeneutics constitute a major departure from these approaches. Thus paralogic theorists posit alternate conceptions of community, knowledge and consensus, discourse conventions, and collaborations—conceptions that would significantly modify research in professional communication done under a paralogic rubric.

Community

Paralogic hermeneutics' most fundamental critique of social construction involves its concept of community. According to paralogic theorists, this constructionist concept leads logically to the claim that the shared bodies of beliefs constituting communities are internalized by community members as ways of organizing experience existing before and mediating understanding. Based on the work of

Davidson, paralogic theorists term these ways of organizing experience *conceptual schemes* (Dasenbrock, 1991, pp. 9-11; Kent, 1991, pp. 425-426, this volume, in press-a, in press-b).

Paralogic theorists claim that the constructionist concept of community, which appears to entail conceptual schemes, cannot account for certain commonsense notions we hold. If, for example, a community's socially justified beliefs form a conceptual scheme or framework endorsed by the community, then beliefs must be entirely relative to communities, and no belief can be said to be truer than another (Dasenbrock, 1991, p. 10; Kent, 1991, p. 426). In addition, if beliefs as conceptual schemes are essential because they enable people to understand one another, then members of a community can only know the unique conceptual schemes that shape their lives. Paralogic theorists assert that these logical extensions of social construction are problematic because we clearly do hold statements to be true and we clearly do understand others who live in different communities (Dasenbrock, 1991, pp. 10-12; Kent, 1991, p. 426, in press-b). Thus, citing Davidson, these theorists claim that the social constructionist understanding of community as an entity defined by a conceptual scheme must simply be incorrect (Dasenbrock, 1991, pp. 10-11; Kent, 1991, p. 428, this volume). At the same time, the ideologic understanding of community, where its norms would be broadened to include marginalized voices, is also incorrect (Kent, 1991, pp. 439-441). Instead, in order even to recognize communities and understand their discourse, communicants must hold some views in common—views that cannot be relative to the conceptual schemes of the communities to which communicants belong.

Paralogic theorists, then, object to "the very idea" of the constructionist discourse community (Kent, 1991), focusing instead on the rapport experienced by communicants as they interact. This paralogic reformulation of the concept of community, with its insistence on the primacy of communicative interaction, has major implications for professional communication research. Specifically, because paralogic theorists do not believe that communities' conceptual schemes regulate discourse in advance of an interaction, research would no longer examine the ways communities determine communication by means of their shared values, beliefs, and norms. Instead, researchers would focus on the uncodifiable nature of communicative interactions. Paralogic theory also mandates that researchers abandon attempts to make binding generalizations across interactions or arrive at totalizing theories about communities, such as Bruffee's constructionist theory of communities as peers who share socially justified beliefs or ideologic critics' theory of reproductive ideology. To paralogic theorists, all such attempts at total-izing generalizations serve to codify and reduce to a system a herme-neutic act that is essentially uncodifiable. Hence, although descriptions of the way interactions occurred in a management consulting firm or an academic discipline would be possible and even helpful, such descrip-tions could not be used to derive theories that might then be applied to explain other interactions.

Finally, paralogic theory would resist research that attempts to arrive at definitive rules for effective communication within social groups. Basing this resistance again on the uncodifiable nature of the herme-neutic act, paralogic research might offer suggestions about what ap-peared to be useful strategies within specific interactions—what Kent (this volume) terms background knowledge about writing—but para-logic theory would preclude codifying these suggestions as requirements for good writing that could be applied before communicative interac-tions. This focus on communicative interaction and the uncodifiable nature of the hermeneutic act also informs the paralogic concept of knowledge and consensus.

Knowledge and Consensus

Paralogic theorists also disagree with social construction's under-standing of knowledge and consensus, claiming that knowledge cannot, as social construction appears to believe, be viewed as consensually held values, beliefs and norms internalized as community members' concep-tual schemes. In place of this constructionist understanding, paralogic theorists assert that knowledge is an agreement reached with other com-municants through the process of interacting.

In order to reach this agreement, communicants participate in herme-neutic guessing (Kent, 1989b, p. 29), until they arrive at an admittedly temporary rapport (Dasenbrock, 1991, p. 13). This rapport is due to com-municants' adjusting two types of guesses or "theories": what Davidson (1986a) calls prior and passing. As paralogic theorists, using these con-cepts, explain, prior theories are the beginning points for interactions, are "set[s] of assumptions about the dispositions, beliefs, and language-use" held by communicants who are interacting (Dasenbrock, 1991, p. 13). Passing theories are then adjustments to prior theories or guesses arrived at as communicants interact (Dasenbrock, 1991, p. 13; Kent, this volume, in press-b). In Kent's (in press-b) words:

As a speaker speaks and a listener listens, they both possess prior theories
that undergo modification as they speak and listen. As they guess about the
meaning of one another's sentences, they together arrive at a passing
theory, a unique hermeneutic strategy, that will enable them to understand
one another in their own singular situation.

When communicants come to share a passing theory, they have understood one another and achieved the paralogic version of consensus. This
understanding is, however, always temporary as additional interactions
will lead to further adjustments in both prior and passing theories.

As was true with the concept of community, this paralogic concept
of knowledge and consensus, with its stress on the uncodifiable nature
of the process by which passing theories are generated, has implications
for research in professional communication. Specifically, research informed by this approach would redirect the constructionist and ideologic focus on the means by which beliefs are socially validated and
incorporated into a community's knowledge store—in ideologic terms,
the community's tendency to reproduce itself. Instead of this stress on
knowledge as communities' social justification of belief, paralogic research would focus on the open-ended and dialogic character of the
process of reaching temporary agreement, as communicants adjust their
hermeneutic guesses. At issue would be the predisposition for constructionists and ideologic theorists to view the community as determining
a body of knowledge, either through its consensually held norms or
through its repressive authority structures, a view that, for paralogic
theorists, violates the indeterminate nature of the hermeneutic act.

In addition, paralogic theorists would again argue with research that
attempts to arrive at generalizations, this time concerning social justification. By implication, then, research would no longer aim at theories,
such as Spilka's (1990) on the role of orality or Winsor's (1989) on inscription, about the way in which communication enables consensus.
Nor would research aim at theories, such as Trimbur's (1989), that
describe how consensus making can be improved. Because the interpretive process cannot be codified in advance of an interaction, no such
generalization or theory can ensure that communicants will, in fact,
reach agreement. Paralogic research thus would insist that the ability
of communicants to converge on passing theories can only be described,
never systematized in advance of communicative interactions by generalizations about knowledge and consensus. Although such descriptions
might well provide interesting and useful stories, adding to our back-

ground understanding of writing and enabling us to better arrive at effective passing theories, such background understanding should never be confused with prior rules governing communicative interaction. This same focus, which insists on the open-ended, uncodifiable nature of the hermeneutic act, informs the paralogic view of discourse conventions.

Discourse Conventions

According to paralogic theorists, the constructionist and ideologic tenet that discourse conventions are entirely relative to communities makes impossible an explanation of how we can communicate (Kent, in press-a). Thus paralogic theorists do not subscribe to what they identify as the constructionist and ideologic position: that discourse conventions are prior requirements for communication, forming a codfiable system that can be mastered and known only by community members.

In place of this constructionist and ideologic position, paralogic theorists claim that discourse conventions derive meaning from their use by communicants (Kent, 1989b). More precisely, discourse conventions are important as parts of the matching process by which communicants attempt to suit their beliefs about language to the beliefs of others (Kent, in press-a), thus enabling communicants who share such beliefs to reach agreement.

For example, when communicants share beliefs about such conventions as the imperative verbs found in sets of instructions, sexist or nonsexist language practices, visual cues such as boldface type or indentation and so on, these beliefs require communicants to make "fewer guesses" (Kent, 1991, p. 433) about each others' interpretations. Thus discourse conventions facilitate, but are not prior requirements for, the interpretive act. Moreover, because discourse production and reception escape codification, knowing or using discourse conventions can never be equated in advance with effective communication, but only with effective use of background knowledge (Kent, this volume) and perceptive hermeneutic guessing in particular interactions.

In a study of realist ethnographies, Kent (in press-c) illustrates this paralogic view of discourse conventions by describing the way in which two conventions —the objectification of narration and of data—enable communicants' beliefs about the ethnographies they read to cohere with their previously held beliefs about the way in which realist ethnographies in general are written. In Kent's analysis, the objectification of narration and of data are integral to the hermeneutic process by which communicants interpret ethnographies and agree on meaning.

Kent's study—one of the few in professional communication done under the rubric of paralogic hermeneutics—is important in considering the direction for such research on discourse conventions. For example, paralogic theory would reevaluate constructionist research focusing on discourse conventions as constituting communities and ideology-based research on discourse conventions as vehicles of power and control. Taking communicative interactions as its starting point, studies informed by paralogic theory would focus on conventions as they facilitate, or do not facilitate, specific hermeneutic acts but not on conventions as prior to and dominating interactions.

At the same time, paralogic theory would urge researchers to resist both social construction's tendency to equate effective writing with the conventions of communities and ideologic critics' tendency to valorize resistance to the conventions of dominant interests as the means to more heterogeneous—and thus better—discourse. Although researchers might still describe effective communicative interactions, the uncodifiability of the hermeneutic act would prevent equating that effectiveness with theories about what constitutes good writing.

This reformulation of the social constructionist and ideologic concepts of community, knowledge and consensus, and discourse conventions, in light of paralogic theory, also leads to a reformulation of collaboration, as the basis of communicative interaction and as a pedagogic strategy.

Collaboration

Because paralogic theorists posit that the hermeneutic act is dialogic and hence essentially collaborative (Kent, 1989b, pp. 36-37), they would accept collaborative pedagogy as possibly useful in the writing classroom. Paralogic theory, however, does challenge the theoretical underpinnings as well as the aims of much current research on collaborative pedagogy. At issue is social constructionist and ideologic theorists' implicit claim that collaborative groups and strategies can be used to teach communicative interactions and thus either help students produce good writing or empower them as individuals.

From a paralogic perspective, the idea that collaboration can be used to teach communicative interactions is based on the mistaken notion that communication is a codifiable system. When Bruffee, for example, argues that collaboration can be used to acculturate students to discourse groups, he appears to assume that, because communities and conventions are prior requirements for communication, good writing consists of mastering normal discourse. Similarly, when Trimbur or Berlin argue that collaboration (vis-à-vis dissensus or conflict) can help empower students by drawing their attention to the systems of power that shape their lives, these ideologic critics assume that structures of authority regulate communicative interactions.

In both instances, collaborative pedagogy is based on the notion that communities and conventions make communication and structures of authority possible and that collaborative strategies and group work can facilitate either mastery of or resistance to these conventions. For paralogic theorists, however, the hermeneutic act should be the focus of writing pedagogy and the goal of the collaborative classroom. According to Kent (in press-a), writing teachers should "encourage students to think about writing as a communicative interaction and not as a skill, like riding a bike, that can be mastered and internalized." To facilitate this understanding, interaction in the classroom would take the form of one-on-one student/teacher collaborations (Kent, this volume). As collaborator, Kent (in press-a) explains, "the teacher would hold no privileged body of knowledge that the student would need to ferret out. Instead the teacher would collaborate actively with the student to learn to write as we [fellow language users] write."

Collaborative pedagogy informed by paralogic hermeneutics would significantly modify collaboration as it is currently conceptualized in social constructionist and ideologic research and instruction, requiring that teachers and researchers rethink the aims of many collaborative activities as well as the claims of much existing research on collaboration. Paralogic hermeneutics, for example, would require that teachers of professional communication no longer hold acculturation as the aim of peer editing groups and small-group writing projects because acculturation assumes that good writing can be reduced to internalizing discourse conventions. Paralogic theorists argue that collaborative group members cannot draw on a store of conventions that will necessarily translate into effective documents. Conventions are background knowledge or strategies that group members might find useful, but the effectiveness of these conventions cannot be prescribed in advance.

Because strategies for effective communication cannot be known before, but only emerge during, communicative interactions, paralogic hermeneutics also asks that researchers not attempt to codify collaborative

interchanges and generalize about the relationship between these interchanges and effective or ineffective writing. For paralogic theorists,
discoveries about interpersonal exchanges that emerge from empirical
studies or observations of collaborative groups might be helpful in
expanding students' fund of background knowledge to be drawn on
when writing, but these discoveries cannot be codified to determine in
advance that which will be effective for subsequent groups.

Finally, paralogic theory would ask that ideologic critics rethink the
issues of collaboration and empowerment. Because Myers, Berlin, and
Trimbur see structures of authority, like they see conventions, as forces
that exist before communicative interaction, their understanding of
dissensus and other such collaborative strategies can never truly empower individuals. Collaborative groups still remain imprisoned within
norms that constrain communication. Thus, although paralogic theorists
would agree with many of the political questions ideologic critics raise
about collaboration and empowerment, paralogic theorists would embed
issues of power within specific communicative interactions, making
empowerment a possibility because communicants are no longer prisoners of prior conceptual schemes. Instead, communicants can be empowered through their sensitivity to power relations in the interpretive
process.

In summary, paralogic hermeneutics represents a major departure from
the social constructionist and ideologic approaches. Paralogic theorists,
for example, differ from social constructionist and ideologic theorists
on such fundamental issues as the nature of both interpretation and communication. These differences, in turn, lead paralogic theorists to rein-
terpret the constructionist and ideologic concepts of community, knowledge and consensus, discourse conventions, and collaboration, in light
of assertions about the external nature and the importance of communicative interaction and the uncodifiability of the interpretive act.

The existence of these three approaches—all social in nature but at
times radically different in their theoretical presuppositions—under-
scores our claim for considerable diversity among theorists adopting a
social perspective. In the following section, we assess the major contributions of each approach and we speculate on the challenges the ideologic and paralogic hermeneutic approaches are posing, in terms of the
directions for further research.

Assessment and Directions
for Future Research

Any assessment of the three approaches must rightly begin by ac-knowledging the major contribution that constructionist theory has made to professional communication studies. Without question, Bruffee and others attempting to articulate a constructionist position deserve credit for bringing attention to key presuppositions in social theory and for showing how these presuppositions might recast conversations about writing pedagogy and research. As many of the studies in profes-sional communication that we cite in this chapter suggest, construction-ist notions of knowledge and communication as socially situated phe-nomena have effected a major shift in many researchers' investigations toward the culturally specific and the local contexts of communication. The result has been a widening of the research agenda in professional communication, to include qualitative and quantitative studies of the organizational, institutional, and classroom contexts in which writing occurs.

Constructionist theory might also be credited with helping establish professional communication studies within the larger arena of cultural studies. Because many scholars in professional communication now share with scholars in literature and the social sciences a belief in the constructionist concept of the social nature of communication, con-structionist theory has helped foster links between professional com-munication and other disciplines. The result has been to bolster an identity for professional communication as a cross-disciplinary field.

The ideologic and paralogic hermeneutic approaches, meanwhile, must be credited with reinvigorating discussions of social theory and with deepening our understanding of communication as socially based. Both approaches have pointed out the dangers of uncritically accepting the presuppositions and vocabulary derived from constructionist theory. By working to unpack expressions like *discourse communities, com-munity of knowledgeable peers,* and *the social construction of knowl-edge,* ideologic and paralogic critics have encouraged more careful scrutiny of constructionist claims and more thorough consideration of the implications of these claims.

In so doing, both the ideologic and paralogic approaches offer im-portant and perhaps even dramatic challenges for the direction of pro-fessional communication research. At the heart of these challenges, moreover, is the essential question: Can the ideologic and paralogic hermeneutic approaches be incorporated into existing research agendas and professional communication courses without undercutting the way in which many researchers and teachers in professional communication currently define and conduct their work?

Regarding the ideologic approach, the answer to this question is both yes and no. The answer is yes if we take the ideologic approach to mean enlightening students about some issues of gender, race, and class and the tendency for certain voices to be marginalized in discourse situations. Researchers like Lay, for example, give every indication that communication courses can incorporate instruction on gender bias and that this instruction can translate into larger cultural change, as students take more egalitarian communication strategies to their collaborations in the workplace.

The answer is no, however—the ideologic approach cannot be incorporated into existing ways researchers and teachers in professional communication define and conduct their work—if the ideologic approach is understood to mean critiquing and maybe even resisting the larger economic values of a commodity culture. Such critiques could significantly refocus research and pedagogy in ways that threaten the idea of a cooperative relationship existing between academe and industry and between research and professional practice. For example, by refocusing research on communication as a vehicle for serving and protecting the economic interests of dominant groups, the ideologic approach creates a potentially adversarial relationship between researchers and the very organizations and agencies that are the object of study or the sources of funding for researchers' work. This adversarial relationship subsequently calls into question the likelihood of researchers gaining access to organizations for ethnographic or empirical study as well as the opportunities for researchers to pursue questions that may be contrary to the political interests of funding agencies. Pedagogy informed by an ideologic approach could similarly threaten the possibilities for cooperation by undercutting the idea that writing in the classroom and writing in the workplace are complementary, mutually supportive activities. Rather than helping students make the transition to the communication demands of their jobs, the ideologic approach would help students understand the ways in which workplace communication practices may protect private interests and subvert the larger public good.

Because the paralogic hermeneutic approach challenges existing ideas about the codifiability of communicative processes, the approach, like the ideologic, refocuses research and pedagogy in ways seemingly incompatible with current views in professional communication. For example, the paralogic concept of communication—that communication is enabled through individual interactions rather than by communities and conventions as prior conceptual schemes—undercuts an implicit goal of many studies of workplace writing: to identify and codify effective writing practices. Denying that ethnographic or empirical studies can allow us to arrive at generalizations about the writing process, paralogic theory thus recasts research claims, holding that such studies may yield rich and interesting stories but cannot lead to larger theories about collaborative or other writing processes in the workplace.

Paralogic hermeneutics effects an even more radical shift in the pedagogic goals of professional communication courses. Again, by refocusing attention on communicative interactions and the uncodifiability of these interactions, paralogic theory undercuts traditional ideas about writing as a teachable skill. The approach thus calls for a dramatic reinvisioning of the function of professional writing classes, including the instructional techniques employed there, away from either enculturation into or resistance to the norms of communities and toward a concept more akin to one-to-one collaboration.

In light of the provoking challenges that the ideologic and paralogic hermeneutic approaches offer social conceptions of writing research and pedagogy, it is difficult to say the extent to which future research will actually be driven by either of these approaches. Clearly these approaches expand our understanding of a social perspective and open new possibilities for conceptualizing the social nature of our work. Because the ideologic and paralogic hermeneutic approaches also, however, challenge certain fundamental beliefs about our research and teaching, we can only wait to see the nature of the debate and discussion that these approaches may generate and whether they will, in fact, lead us to a more self-conscious examination of the social perspective in professional communication.

Notes

1. Because Bruffee's articulation of social construction appears to be the most commonly cited in professional communication research, we focus here on Bruffee's

interpretation of constructionists tenets. Clearly, some critics have taken issue with Bruffee's readings of these sources (see, for example, Greene, 1990).

2. It is not possible here to work out the full line of reasoning behind the paralogic hermeneutic critique of social construction and internalism. Much of this reasoning builds from such key concepts as *conceptual schemes* and *passing theories* (which we touch on in our subsequent discussion)—concepts derived directly from Davidson's philosophy of language. For additional background, see Davidson (1986a, 1986b, 1986c, 1986d).

2

Rhetoric Unbound:
Discourse, Community, and Knowledge

BRUCE HERZBERG

Rhetoric, Plato warned long ago, is a dangerous thing. To understand his fears, we need only consider the many examples of orators who have used their powers to foment hatred, incite to genocide, and pervert justice in a thousand ways. These evils of rhetoric come, Plato argues, from the principle, advocated by the Sophists, of substituting persuasion for truth. Could intelligent, responsible people actually propose such a monstrous principle? The way Plato makes his case, the Sophists seem neither intelligent nor responsible. But Plato's formulation of the premise of Sophistic rhetoric is tendentious, to say the least: The idea that rhetoric *substitutes* persuasion for truth implies that the truth can simply be ignored. The Sophists make a rather different and more radical claim, namely that there is no such truth as Plato imagines, only the truths that operate, however locally and temporarily, for a given group of people. Rhetoric, in the Sophists' view, is the means by which such truths come to be accepted and through which they then become effective. The Sophists regard all knowledge as probabilistic, contingent, and uncertain, and all demonstrations that purport to reveal the truth as persuasions that lead to a conviction of truth.

Plato, shocked by this radical view, labored to separate truth, which concerned things that might be known, from knowledge, which, being human, might well be distorted or partial. Rhetoric, in Plato's view, contributes to the distortion of truth and the degradation of knowledge precisely because it takes a probabilistic rather than an absolute view of truth itself.

In his dialogues, Plato treats the Sophists as quibblers, concerned only with winning arguments by any means necessary. They were not, to be sure, seekers of truth in the Platonic sense. Plato's arguments against them were so successful that it was not until the 19th century that the Sophists were regarded as anything other than the corrupters of Greek youth or self-serving hypocrites. Plato's victory seemed justified and complete. Nowadays, however, the hero of the story of Plato and the Sophists is no longer so self-evident. The underdogs have won a great deal of sympathy, for the views of the Sophists adumbrate our own.

A social perspective on communication might well mean little more than a special concern for audience. But the sort of social perspective taken by the Sophists—and a few hardy souls who followed their lead —provides a distinctive view of the interdependencies of rhetoric, communities, and knowledge. In this chapter, I examine the Sophists' ideas and then present a few of the very rare instances between the Sophists' time and our own when similar views have been expressed. Giambattista Vico, George Campbell, and Friedrich Nietzsche, whose ideas I will be looking at, raised again the Sophistic suspicion that access to the truth may not be possible, that what we take to be true may be a communal and rhetorical construction of beliefs, conventions, and rationalizations.

The Sophists

Foucault (1971/1981), in his lecture "The Order of Discourse," tells this fable about the origin of philosophy and the division it drew between truth and discourse:

> For the Greek poets of the sixth century BC, the true discourse . . . was the one pronounced by men who spoke as of right and according to the required ritual; the discourse which dispensed justice and gave everyone his share; the discourse which in prophesying the future not only announced what was going to happen but helped to make it happen. . . . Yet already a century later the highest truth no longer resided in what discourse was or

did, but in what it said: a day came when truth was displaced from the ritu-
alized, efficacious and just act of enunciation towards the utterance itself,
its meaning, its form, its object and its relation to its reference. Between
Hesiod and Plato a certain division was established, separating true discourse from false discourse: a new division because henceforth the true
discourse is no longer precious and desirable, since it is no longer the one
linked to the exercise of power. The Sophist is banished. (p. 54)

Between Hesiod and Plato, Foucault tells us, the magical quality of
language is dispelled, the essential similitude between signs and their
referents is disrupted. Discourse begins to disappear, to become invisible,
its workings hidden by new relations of knowledge that placed the truth
outside of language. The disappearance of discourse takes time. At many
other moments in history Foucault (1971/1981) finds "new forms of the
will to truth. . . . From the great Platonic division, the will to truth had its
own history" (pp. 54-55).

The will to truth is manifest in attempts to define truth in such a way
as to make it knowable but not rhetorical. As Foucault suggests here,
there have been many such manifestations, including Platonic idealism,
the revealed truth of religion, and the empirical truth of science. Rorty
(1979), speaking in a similar vein, points out that philosophy has always
been devoted to "the construction of a permanent, neutral framework
for inquiry, and thus for all of culture" (p. 8). There are many ways to
imagine this framework and its location—it may be in the world itself,
in the mechanism of perception, in the mind, or even in language. The
key terms in Rorty's formulation are *neutral,* meaning not influenced
by belief, and *culture,* meaning that the foundations of knowledge are
understood to precede other kinds of human activity.

The "true" discourse of the Sophists—to return to Foucault's charac-
terization—embodies and creates truth because it speaks to persuade,
to decide, to wield political power, to do justice, and to preserve cultural
cohesion. But after Plato, after Sophistry is defeated by philosophy,
truth is displaced from the social to the ideal. The proper role of discourse in Plato's scheme is to *represent* truth. The truth is to be found,
not created. Truth in this nondiscursive sense can be known without
being said; statements are linguistic phenomena (they have form and
meaning, as Foucault says) but it is what they express (their object or
reference) that may be judged as true or false. Discourse always conveys
something other than itself, in the Platonic view. It is not the act of
discourse but the signified object of discourse that determines whether

the discourse is true or not. Thus, for Plato and Aristotle, true philosophers engaged in dialectic to find a truth that was beyond speech, a truth that could thereafter be conveyed persuasively in rhetoric to a nonphilosophical audience. What Foucault calls "the will to truth" is the effort to define and secure the boundaries of a truth that can be found and kept. And one consequence of this effort has been to isolate rhetoric (and its realm of probability, argument, and persuasion) in order to keep it from contaminating the truth.

The Sophists believed that reality was unknowable. The world existed, to be sure, but it was what people made of it that constituted reality. The search for something more real than that, some truth that transcended human experience, led only to the speculations of philosophers, whose opinions shift back and forth constantly (as the Sophist Gorgias says in his *Helen*), blown by the uncertain winds of the moment's opinion. The Sophists argued that knowledge was necessarily uncertain. Similarly, social values were consensual—what was good and right for the Athenians might not be so for the Spartans. The uncertainty of knowledge and the relativity of values did not, however, mean chaos and anarchy. To the contrary: The Sophists concluded from these premises that virtue, being a set of social conventions, could be taught. The radical aspect of this conclusion was that virtue could be taught to anyone—nobody was predisposed by birth to be more cultured or virtuous. To conservatives like Plato, the Sophists appeared to be abandoning morality, which was founded on divinely given standards of conduct; they were certainly abandoning the search for truth, truth that was unequivocal and universal; and they seemed to be sowing the seeds of anarchy as well, for they disputed the notion that class privilege was founded on the divine gift of virtue.

The Sophists had a complex sense of the power of language. In the democratic experiments of the 5th century B.C. in Greece, when public discourse led to public policy and justice was dispensed in response to the arguments of litigants, the ability of language to move and persuade audiences was not by any means limited to epic, lyric, and tragedy. The Sophists, in examining the power of public discourse, were theorizers not only of democracy but of the ways that the power of poetry and magic pervaded all language.

In *Helen* (c. 414 B.C./1990), Gorgias defines persuasive speech as poetry without meter and describes the effects of rhetoric as magical. *Helen* is a show speech in which Gorgias demonstrates his art and discusses some of its premises. He begins by asking if Helen was at all to blame for going off with Paris and thereby precipitating the Trojan war. Gorgias defends Helen by suggesting that there are four possible reasons for her behavior: the gods, force, speech, or love. As for the first possible cause, if "by fate and gods the cause had been decreed, Helen must of all disgrace be freed" (p. 34). She is, of course, not culpable if Paris took her away by force. Love, says Gorgias, has power even over the gods, so if Helen fell in love with Paris it was misfortune but no crime. And speech, says Gorgias, "is a powerful lord that with the smallest and most invisible body accomplishes most god-like works. It can banish fear and remove grief and instill pleasure and enhance pity" (p. 35). Speech can cause fear, pity, and "tearful longing." Speech can induce pleasure and remove pain. Speech is incantation, witchcraft, magic—it beguiles and persuades. Speech works on the soul, says Gorgias, like drugs on the body. Who could resist? If Paris used rhetoric, Helen is still exonerated.

The belief in the divine origin and magic power of speech is repeated in many ancient cultures—in the Bible story of creation and in Adam's naming the animals, in the story (told by Plato, c. 360 B.C./1971, in the *Cratylus*) of the ancient "legislator" who did Adam's business of naming for the ancestors of the Greeks, in the story (told by Plato, c.370 B.C./1985b, in the *Phaedrus*) about the gift of writing made by the Egyptian god Theuth. Belief in the magical efficacy of speech persists in praying and cursing, taboo words, magic formulas, secret and unutterable names of people and gods, and belief in the special powers of a culture's originary language. In ancient Greece, that power descended through poetry and tragedy to the Sophists' notion of rhetoric.

The Sophists seem generally to be skeptics and iconoclasts. *Helen* is typical of their desire to challenge traditional religious and historical assumptions. They were not engaged, it would appear, in magical thinking. The magic of speech is an engaging expression of the belief that language has powerful psychological effects. (The Sophist Prodicus, c. 450 B.C./1992, argued that religion came about precisely because people attributed natural forces to the gods or elevated to divinity those people who invented useful techniques in, for example, agriculture. So too with language: Perhaps the attribution of speech to the gods followed the recognition of the poetic or even rhetorical effectiveness of language.)

The power of language, for the Sophists, is not limited to pity, fear, longing, or other emotions. It extends to knowledge. In the epistemology of the Sophists, as we have seen, there is no knowledge of transcendent truth. The Sophist Protagoras (c.450 B.C./1972) says, "Of all things the measure is man, of things that are that they are, and of things that are not that they are not" (p. 18). The anonymous Sophistic tract *Dissoi Logoi* (c. 420 B.C./1972) takes this view in the social sphere, arguing that notions of morality and justice vary from place to place. What counts as right in one time and place may not be right for another. Thus the Sophists maintained the doctrines of *kairos,* or "timeliness," and *to prepon,* or "propriety."

As Foucault (1971/1981) says so cautiously, "between Hesiod and Plato a certain division was established" between Sophistic rhetoric and the Platonic "will to truth" (p. 54). The great rhetoric teacher Isocrates (c.390 B.C./1929) wrote the tract "Against the Sophists" to advertise the pragmatic principles of his school and distance himself from the Sophists' tendency toward stylistic display and from their excessive claims about the teaching of virtue. The principle of *kairos* was, for Isocrates, a motto pointing toward careful consideration of details before rendering a moral judgment: Virtue itself was a function of action, of actually making sound moral judgments. In softening the claims of the Sophists and focusing on the practice of rhetoric in the realm of public affairs, Isocrates agrees in some measure with his contemporary Plato, who attacks the Sophists in his dialogue *Gorgias* (1985a) and argues for limits on rhetoric in his *Phaedrus* (1985b).

Plato is afraid that education centered on rhetoric will be attractive because it can produce persuasive and therefore powerful orators without necessarily teaching any substantial knowledge. A Sophistic orator can presumably create belief in any proposition that seems plausible and feel no obligation to search for the truth. In *Gorgias,* Plato (c.386 B.C/1985a) rejects rhetoric outright, urging that it is immoral and dangerous. Taking up the very images offered by Gorgias in *Helen* (c. 414 B.C./1990), Plato warned that rhetoric was like drugs or magic and therefore not to be trusted. Plato did not believe that the genie of rhetoric could be stuffed back into its bottle. He hoped, at first, to denigrate rhetoric so that audiences would shun the self-proclaimed rhetors and students avoid their schools. In *Phaedrus,* Plato (c.370 B.C./1985b) moderates his position, calling on rhetoric to communicate knowledge and to use its persuasive power to bring the truth to various kinds of audiences. Here he approaches the position taken by Isocrates.

The arguments of Isocrates, of Plato in the *Phaedrus,* and of Plato's student Aristotle prevailed in this extended debate. Rhetoric remained at the center of education in the ancient world, but it took on a practical or managerial role. Rhetoric thus conceived is bound up with forms for constructing discourse that will be effective in conveying information or swaying belief. Managerial rhetoric specifies methods for producing discourse, for organizing it, for finding arguments, for appealing to audiences, and so on. The discipline of rhetoric, firmly established by Aristotle's elaborate system, could be distinguished clearly from other fields of inquiry. Thus philosophy, with its independent existence, could pursue the truth, which rhetoric could draw on as needed.

In Aristotle's (c.323 B.C./1984) *Rhetoric,* the concept of audience helps to damp down the epistemic problem that arises if we believe that people perceive the world differently. In Aristotle's system, people are convinced by logic, which is universal, and by emotion, which is fairly uniform. Identify the people in the audience; note their susceptibility to logic (their objectivity and intelligence); add appeals appropriate to their gender, age, nationality, and economic and social class; and there you have it: Conviction can be created by taking account (not to say advantage) of social circumstances.

Plato did admit, finally, that some knowledge was necessarily probabilistic, and that rhetoric might be useful in sorting out such cases. Aristotle codified these cases (of law, policy, and ceremony) in such a way as to retain the distinction between what might be known with certainty and what might be known only probabilistically. The epistemic role of rhetoric was in this way tightly controlled and made to seem a method for handling special cases.

Francis Bacon and the Challenge
of Epistemology

In Christian Europe during the Middle Ages, dialectic served as the discursive medium for seeking the truth. Dialectic was held to be a strictly logical method for constructing and criticizing arguments. There were many forms of dialectic—including propositional logic, the study of grammar and semantics, syllogistic reasoning (subdivided into many types), and a form of disputation that worked by thesis and antithesis. In the later Middle Ages and in the Renaissance, many thinkers found fault with the limitations of dialectic—which, among other things, excluded as false any proposition or conclusion that conflicted with divine revelation. But it was not until quite late in the Renaissance that dialectical logic was seriously challenged by Bacon, the champion of empirical science. Syllogistic logic with its deductive approach to knowledge simply could not, Bacon (1620/1955) declared, discover anything new. Inquiry, observation, and inductive reasoning—empiricism, in short—was the real foundation of knowledge.

Bacon does not, to be sure, offer anything like a social view of knowledge. Nor does he recuperate Sophistic accounts of rhetoric. Yet in his efforts to establish the grounds of empiricism, Bacon needed to address the question of how people acquire knowledge. If we are to trust our perceptions, generalizations, and conclusions, we need to understand the psychology of knowing. Is the mechanism of knowledge so uniform that we can assume that what we know is true for everyone? Even Plato had argued that social condition could affect one's knowledge. The lower orders, he said, were absorbed in mechanical labors that conditioned the mind as well as the body; hence they were not fit to pursue higher learning. Bacon, too, acknowledged that there were a number of kinds of impediments to understanding. These he called the *idols*.

The Idols of the Tribe are "natural" flaws in our ability to interpret what we perceive: "The human understanding is like a false mirror, which, receiving rays irregularly, distorts and discolors the nature of things by mingling its own nature with it" (Bacon, 1620/1955, p. 470). Can anything be done about such distortions? Yes, indeed: Bacon (1620/1955) prefaces his discussion of the idols with the assertion that "the formation of ideas and axioms by true induction is no doubt the proper remedy to be applied for the keeping off and clearing away of idols" (p. 469). Yet there are so many idols—those of personal predilections, those of social and political orientation, those imposed by philosophy and religion—that in the very act of naming them, Bacon raises the specter of their possible intransigence, for how can we be sure that we have kept them off or cleared them away?

Epistemology becomes, from this moment, the central concern of philosophy. Although Bacon seems sanguine about the triumph of empiricism, it was no easy task to identify, for example, the clear and distinct ideas that would be the basis of inductive thinking, or even to determine what inductive processes could be assumed to be universal. The notion of "clearing away" the sources of distortion was a way of attributing at least some of these difficulties to a malign "outside" influence. Rhetoric, and even language itself, were often cast in this role. Bacon himself complains that far too many words have multiple denotations and connotations. Bacon's followers vigorously sought language reforms to remedy this situation, hoping to fix definitions, establish rules of grammar, and promote a plain style in opposition to the extravagances associated with rhetoric.

Giambattista Vico:
Rhetoric and Rationalism

With rare exceptions, both rationalists and empiricists believed that cultural differences among societies and individuals were simply sources of error: Such differences constituted neither different systems of knowing nor different epistemologies. When Vico, an Italian professor of rhetoric, attacked Cartesianism, arguing that it was indeed a cultural phenomenon and therefore just another system of knowledge—no better than and in some ways inferior to others—he was regarded as a reactionary. Science was on the march, it seemed—it just needed to work out some kinks.

Vico objected not to science itself, but to Descartes's claim that any knowledge that was merely probable was not knowledge at all. Vico feared that if contingent knowledge were not respected, then civic virtue as well as many "merely probable" fields of knowledge like law, history, and medicine, were threatened. "Human events," Vico (1709/1965) warned, " are dominated by Chance and Choice" (p. 34). Those trained in the mind-set encouraged by Cartesianism, which denigrates argument and persuasion, may wind up unsuited for "prudential behavior in life" (p. 34). The argument between Vico and the Cartesians inverts the argument between Plato and the Sophists in an interesting way. Vico argues that the search for certainty is a limited concern, not an overarching one. Cartesian rationalism, with its focus on linear causality, must be confined to the study of science and not allowed to affect the rest of the curriculum, where it might dangerously interfere with ethical training. It is, to the contrary, the study of rhetoric, with its concern for probable but not absolute truths, its sensitivity to the multiple causes for many human events, and its awareness of change, that provides the most sensible approach to the wide range of human knowledge.

Vico worked out his notion of epistemological relativism in a number of ways. He took, for example, what we might call a phenomenological approach to history, arguing that it was necessary to reconstruct the consciousness of a time and place in order to understand it. Language analysis is particularly important in this effort, according to Vico, because the form of a language and its etymology are clues to the organization of a society and to the psychology of its citizens. The embedded metaphors of a language speak of the origins of the society (what ideas are built on what prior images), for example. The rationalists' suspicion of language and rhetoric seems particularly reprehensible to Vico, for so much of human reason is bound up in language.

The fullest expression of Vico's (1744/1948) sense of the sociological conditions of knowledge comes in *The New Science*, where he works out the idea of the stages of history. Different eras, he says, have different methods of constructing knowledge and interpreting the world. The particular approach depends on need and circumstance, conditioned by previous experience, the organization of society and the place of the individual within it. Vico posits three general stages of world history. In the first, knowledge is constructed by metaphoric generalization and is focused on the need to establish basic institutions. In the second stage, national identity comes to the fore and political organizations develop. The final stage sees the self-conscious study of human knowledge as well as greater democracy. Each stage takes something different for granted, sees new problems, and uses language in a different way. Vico is still universalizing, of course, seeking a common genesis of language and a shared experience of world history. Much of Vico's speculation is taken up with looking for a kind of certainty or an alternative way of conceiving of certainty. This certainty is to be found, he says, in the examination of human actions rather than in the mysteries of God's creation. Applying Vico's ideas to Vico himself, we may say that he was able to criticize but not escape his time and place: The idea that some knowledge must be secure and unchanging permeates his analyses. Then again, it would, no doubt, have been imprudent to stand directly in the path of the steamroller of Enlightenment philosophy. Vico hoped, at least in his earlier works, to be persuasive. He was not persuasive, though, until much later, when changing historical circumstances permitted a reexamination of the premises of the new science.

George Campbell, Science, and Psychology

The reaction of most rhetoricians to the triumph of empiricism and science was, on the one hand, to promote a style characterized by *perspi-*

cuity or clarity and, on the other, to link rhetoric with literature and other forms of writing as an art of composition and criticism. Both of these moves sidestepped the issue of epistemology that the Sophists and Vico saw as implications of rhetoric. Rhetoric continued to focus on the organization of various types of discourse, on the judicious use of persuasive strategies aimed at the audience's emotions, and on an elegant but not ornate style. One philosophically minded rhetorician, however, Scottish clergyman Campbell, linked rhetoric with the still-young science of psychology and in so doing brought rhetoric into contact with epistemology.

Campbell bases his analysis of rhetoric on Bacon's model of the psyche, which posits the notion of separable mental faculties. Each faculty is appealed to in a different way. The faculties are the understanding, the imagination, the passions, and the will. Persuasion comes from appealing to the faculties in the proper order. The rhetor must first address the understanding, seeking to inform (through clear description) and convince (through logical argument). But logic touches only a part of us: The imagination must be stirred by art, the passions moved by pathos, and finally the will must be stimulated by a powerful call to action. In his discussion of logic and argument, Campbell draws a distinction between scientific and moral reasoning. Like Vico, Campbell notes that the former operates in a rather limited realm, seeking to derive a single inevitable conclusion from a chain of logical links. In disputes about human concerns like ethics and religion, however, there will probably be good evidence on both sides of a case. However, Campbell argues, it is not true that because a case is not subject to scientific demonstration, it is beyond the realm of reason. In the many areas where science does not apply, moral evidence—experience, analogy, testimony, and probability—are brought to bear in orderly ways (Campbell, 1776/1963).

Having reached this point in his presentation, Campbell argues that all of the forms of evidence operate the same way: They produce conviction by steps, founding each proposition on the memory of having been satisfied by the previous step. Both scientific and moral evidence operate this way. In other words, scientific demonstration is not really different in kind from, say, testimony. Think, Campbell says, of how many scientific propositions we believe on the basis of the testimony of authorities. Campbell did not develop this argument any further. Perhaps it was his goal to raise the stature of rhetoric rather than to challenge the epistemological basis of science. For all the very wide currency of his *Philosophy of Rhetoric* (1776/1963), which contains this argument, Campbell has received virtually no notice for this particular contribution.

Friedrich Nietzsche and the
Rhetorical Critique of Philosophy

Late in the 19th century, another professor of rhetoric—one of the few who took both Vico and the Sophists seriously—launched a sustained assault on the philosophers and their will to truth. Nietzsche, professor of rhetoric at the University of Basel from 1872 to 1874, declared that it is a foolish delusion to believe in or search for a neutral language that would do no more than convey knowledge. The very idea that knowledge could exist independent of language is, he says, quite ridiculous, an arrogant assumption made by an ignorant creature.

To begin with, all language, says Nietzsche (1873/1989), is permeated with and inseparable from rhetoric:

> There is obviously no unrhetorical "naturalness" of language to which one could appeal; language itself is the result of purely rhetorical arts. The power to discover and to make operative that which works and impresses, with respect to each thing, a power which Aristotle calls rhetoric, is at the same time, the essence of language; the latter is based just as little as rhetoric is upon that which is true, upon the essence of things. (p. 21)

The purpose of language, in short, is not to transmit truths but to act, to gain power over the world, to persuade. The idea that language originated in the sheer act of naming that we observed in the Bible and Greek myth is part of the delusion that Nietzsche seeks to explode.

Nietzsche (1873/1989) argues that all language is metaphoric and that all words are *tropes:* "The tropes are not just occasionally added to words but constitute their most proper nature" (p. 25). There is no such thing as a proper meaning that differs from a figurative one. Rather, language bears a synecdochic relationship to the perceptions or sensations it represents. And the sensations themselves are already representations of the things we perceive. In his essay "On Truth and Lies in a Nonmoral Sense," Nietzsche (1873/1979) elaborates this thesis about language and the knowledge it represents. Knowledge can be no better than language itself: Our commonly held notions of truth are, like language, conventions adopted for social cohesion. "It is only by means of forgetfulness" that we come to believe that our language represents things themselves (Nietzsche, 1873/1979, p. 81). In fact, language, understood as rhetorical, "designates the relations of things to men" (p. 82), not things in themselves.

"What then is truth?" asks Nietzsche (1873/1979), and he answers: "A movable host of metaphors, metonymies, and anthropomorphisms: in short, a sum of human relations which have been poetically and rhetorically intensified, transferred, and embellished, and which, after long usage, seem to a people to be fixed, canonical, and binding" (p. 84). Nietzsche is attempting to delineate the psychological processes that lead to the conviction—widespread if not universal—that the truth can be known and communicated. He acknowledges that social constructions of truth are culturally varied (as in this last passage where he refers to the way "a people" adopt their canons of truth), but prefers the more striking psychological differences that can be drawn from the contrast of humans and, say, insects. It is difficult, he says, for people to admit that insects perceive a different world, "and the question of which of these perceptions of the world is the more correct one is quite meaningless, for this would have to have been decided previously in accordance with the criterion of the *correct perception,* which means, in accordance with a criterion which is *not available*" (Nietzsche, 1873/1979, p. 86).

Nietzsche attempted to use psychology to attack traditional epistemology, but that approach proved to be heavily guarded. Scientific psychology sought universals that (as we have seen even in the early years of the scientific revolution) would ground the observations of empirical science and banish Bacon's Idols. In Nietzsche's day, psychology hoped to reveal a nonlinguistic mechanism that would account for language as well as perception. And, as Rorty reminds us, such a mechanism would precede culture as well. Nietzsche, like Vico, fought the tide and was dismissed as a madman. In our own day, Freud and others having shown that culture shapes the individual psyche through language, Nietzsche sounds like a prophet.

Rhetoric, Community, and Knowledge

In our own enlightened age, the skeptical view of truth is increasingly well accepted. Logic, demonstration, science, and truth are, in this view, to be regarded properly as the titles given at any moment by a particular group to the arguments that it finds most convincing, the assumptions that it questions least, and the paradigms that it follows in defining what counts as knowledge. Consequently, logic is indistinguishable from argument and knowledge is inseparable from rhetoric: As in the subversion Plato feared, knowledge is not conveyed but *produced* by rhetoric. Furthermore, knowledge must be inseparable from the group or community, its material circumstances, its historical conditions, its forms of discourse, and its agenda of self-preservation, replication, and extension.

But what of the evils of rhetoric? Must we now accept that the beliefs of racists, warmongers, and unjust societies are true and unassailable, as good as the truths offered by anyone else? Isn't this substituting persuasion for truth? Perelman and Olbrechts-Tyteca (1958/1969) offer an answer to these questions in the conclusion of *The New Rhetoric,* their massive study of argumentation. Many arguments, they point out, rest on the claim that they are self-evidently true or that they are consonant with the natural order or that they transcend history and culture. This claim is both false and dangerous. There is no knowledge that could possibly be "identical in all normally constituted minds, independently of social and historical circumstances" (Perelman & Olbrechts-Tyteca, 1958/1969, p. 510). Therefore, claims of self-evident or transcendent truth serve to obscure the fact that all such "truths" are in fact convictions based on arguments and thereby "withdraw them beyond the realm of discussion and argumentation" (p. 510). Thus, ironically, Plato's philosophy of ideal truth helps to maintain the dangerous belief that some ideas may be absolutely true and others false. Perelman and Olbrechts-Tyteca (1958/1969) reject all claims to absolute truth in favor of argumentation: "It is because of the possibility of argumentation which provides reasons, but not compelling reasons, that it is possible to escape the dilemma: adherence to an objectively and universally valid truth, or recourse to suggestion and violence to secure acceptance for our opinions and decisions" (p. 514). The truth is not found but made; it is never unassailable; its genesis is to be investigated in the arguments that have established it, in the purposes it serves, in the power it confers. This, I believe, is the strongest and most valuable form that the social view of communication can take. The rhetorical view of knowledge that underlies the social perspective in communication is, finally, a critique of the ways that knowledge is created and the purposes for which it is used.

INTERPRETATION

3

Ideology and the Map:
Toward a Postmodern Visual Design Practice

BEN F. BARTON

MARTHALEE S. BARTON

.

Introduction

In their multidisciplinary review of trends in visual representation, Barton and Barton (1989) document a shift in the literature from a positivistic view of visuals as autonomous structures to a view of visuals as embodiments of cultural and disciplinary conventions. Attacks on positivism in the literature on visual representation are, in fact, now endoxal. It is clearly time for the theoretical science of the visual signifier to move on. But move on to what? Information designer Sless (1986) offers an attractive suggestion: He issues a call to go beyond the study of visuals as embodiments of cultural conventions to the study of the ideologies that,

---

AUTHORS' NOTE: The authors gratefully acknowledge their debt to Brenda Sims, who suggested numerous improvements incorporated in the final draft of this chapter.

in turn, inevitably underlie those conventions. This chapter attempts to answer that call. It elects, moreover, to look at ideology in a *critical* sense, adopting the rationale of social theorist Giddens (1987):

> We can only develop a viable notion of ideology if we either treat the notion as a "neutral" concept; or alternatively if we link it specifically with exploitative domination. The first tactic seems to me essentially uninteresting, because then "ideology" is no different from general descriptive terms such as "idea-system" or "symbolic system." I therefore prefer to adopt the second sense. According to such a standpoint, modes of signification are ideological when they are pressed in the service of sectional interests via the use of power. (p. 270)

Thus a critical study of the ideology of visuals focuses on the ways in which visual signification serves to sustain relations of domination. Moreover, as Giddens notes, ideology performs such service with a Janus face —it privileges or legitimates certain meaning systems but at the same time dissimulates the fact of such privileging. This double nature of ideology provides the impetus for many of those interested in cultural criticism. For Marxist critic S. Hall (1982), for example, the ideological model of power entails

> a way of representing the order of things which endow[s] its limiting perspectives with that natural or divine inevitability which makes them appear universal, natural and coterminous with "reality" itself. This movement—towards the winning of a universal validity and legitimacy for accounts of the world which are partial and particular, and towards the grounding of these particular constructions in the taken-for-grantedness of "the real"—is indeed the characteristic and defining mechanism of "the ideological." (p. 65)

The Map as Quintessentially Ideological

The double nature of ideology in the advertising image has been widely recognized, but the very recognition of its dissimulative nature is its own undoing. For to do its work, ideology depends on its dissimulative nature not being recognized. In this sense, advertising is *not* ideological, for the interpretive tradition is not on its side in its claims of innocence (W. Mitchell, 1987, p. 4). What form *is* quintessentially ideological in this sense? What form, in other words, is traditionally viewed as realistically presenting a "neutral" view of reality, as innocent of a productive, meaning-creative practice? Huck Finn makes a persuasive case for the *map:*

> "If we was going so fast we ought to be past Illinois, oughtn't we?"
> "Certainly."
> "Well, we ain't."
> "What's the reason we ain't?"
> "I know by the color. We're right over Illinois yet. And you can see for yourself that Indiana ain't in sight."
> "I wonder what's the matter with you, Huck. You know by the *color?*"
> "Yes, of course I do."
> "What's the color got to do with it?"
> "It's got everything to do with it. Illinois is green, Indiana is pink. You show me any pink down here, if you can. No, sir; it's green."
> "Indiana *pink?* Why, what a lie!"
> "It ain't no lie; I've seen it on the map, and it's pink." (Twain, 1924, p. 23)

Huck clearly reads the map as a factual rather than a semiological system: that Indiana is pink "ain't no lie." For Huck, the color of the state as represented in the map and the color of the state itself must coincide, for the signifier and the signified have a *natural* rather than a *conventional* relationship.

In Borgès's (1972) justly celebrated parable of mapping practice in a mythical empire, *scale* rather than *color* discrepancy between signifier (a map of the empire) and signified (the empire itself) is perceived intolerantly by the College of Cartographers:

> In that Empire, the craft of Cartography attained such Perfection that the Map of a Single province covered the space of an entire City, and the Map of the Empire itself an entire Province. In the course of Time, these Extensive maps were found somehow wanting, and so the College of Cartographers evolved a Map of the Empire that was of the same Scale as the Empire and that coincided with it point for point. (p. 141)

The Imperial Cartographers clearly share Huck's innocent notion of the map as a factual rather than a semiological system. But the map, as Borgès's continuation of the parable dramatizes, is not coterminous with the reality, and attempts to make or conceive it so are doomed to failure. Unfortunately for the professional cartographers, it remained for succeeding, lay

generations to "deconstruct" the innocence of such maps—and on other
than a conceptual level at that:

> Less attentive to the Study of Cartography, succeeding Generations came
> to judge a map of such Magnitude cumbersome, and, not without Irrever-
> ence, they abandoned it to the Rigors of sun and Rain. In the western
> Deserts, tattered Fragments of the Map are still to be found, Sheltering an
> occasional Beast or beggar; in the whole Nation, no other relic is left of the
> Discipline of Geography. (Borgès, 1972, p. 141)

In short, "the map is *not* the territory" (Hayakawa, 1941, p. 31); yet
because it is perceived as such so readily, it is for Borgès the paradigm
of the sign. And although we don't believe that Indiana is pink, and
would expect cartographic representations of the empire to be radically
scaled, our everyday discourse nevertheless reflects our acceptance of
Twain's and Borgès's fictional recognitions of the map as the paradigm
of the sign. Semiotician Marin (1980) frames the point like this: "The
map . . . is exemplarily the sign and the idea one has of it: thus one says
spontaneously and unhesitatingly . . . of a map of Italy 'That's Italy' "
(p. 47; authors' translation). Or, one might add, of a map of Indiana
"That's Indiana" and of a map of the empire "That's the empire." Most
signs are treated less paradigmatically: Thus we spontaneously say
"That's a diagram of the circuit" and not "That's the circuit."

Still, to establish the map as a quintessentially ideological genre does
not necessarily validate its centrality in an article written by and
targeted to information designers. That is, our institutional and intellectual affiliations are not with departments of geography, and we probably
do not feel threatened by the fact that the failure of Borgès's College of
Cartographers to deconstruct the map led to the demise of their discipline. We may, then, find warrant in the perceived general importance of
the map to contemporary information designers, however unconcerned
they may be with ideological issues. Bertin (1983), for example, focuses on the map in his seminal work *Semiology of Graphics*. Note also
Tufte's (1990) privileging of the map as an exemplary genre for those
interested in the visual display of information: "Standards of excellence
for information design are set by *high quality maps*" (p. 35). And elsewhere: "The most extensive data maps, such as the cancer atlas and the
count of the galaxies, place millions of bits of information on a single

page before our eyes. No other method for the display of statistical information is so powerful" (Tufte, 1983, p. 26).

Let us turn to an analysis of the relation between ideology and power in mapping practices—specifically to an analysis of a discursive mode that, in Barthes's (1970) term, *naturalizes* and universalizes a set of practices so that the phenomenon represented appears to be described rather than constructed (p. 129). We first denaturalize the natural by foregrounding the occulted or naturalized ideological component of map discourse; in other words, we access the *doxa,* which theorist Bourdieu (1977) refers to as "the universe of the undiscussed" (p. 168). We then propose a new visual design practice exploiting avant-garde representational devices for denaturalizing the represented phenomenon, e.g., the collage with its emphasis on the heterogeneity and discontinuity of representational format. Although we take the map as paradigmatic, our discussion of a postmodern visual design practice systematically includes noncartographic examples to suggest more fully the relevance of our analysis to other visual genres. Ultimately, the map in particular and, by implication, visual representations in general are seen as complicit with social-control mechanisms inextricably linked to power and authority.

Denaturalization of the Natural

How do we, as critics, denaturalize the natural? The notion of *hegemony,* derived from the Italian Marxist Gramsci (1971), suggests a promising methodology. According to Marxist literary critic Williams (1973), hegemony—a process by which certain definitions of reality attain dominance in a society, rather than a conspiracy on the part of the ruling group and a passive compliance by the dominated ones—operates in a dual mode: Thus in the hegemonic process "certain meanings and practices are chosen for emphasis, certain other meanings and practices are neglected and excluded, . . . reinterpreted, diluted, or put into forms which support or at least do not contradict other elements within the effective dominant culture" (p. 9). Hence we look first at what is emphasized and then at what is excluded or repressed, at what we will term the *rules of inclusion* and then at the *rules of exclusion* underlying the dual mode in which ideology operates.

Rules of Inclusion

Rules of inclusion determine whether something is mapped, what aspects of a thing are mapped, and what representational strategies and devices are used to map those aspects. These rules amount to either explicit or implicit, overt or covert, claims to power. An explicit or overt link between power and ideology in mapping can be seen in the indispensability of maps for the conduct of war. According to Emmerson (1984): "Making war meant making maps. The National Geographic Society made them in unprecedented numbers, nearly twenty million in 1941-1942" (p. 7). We are not surprised, therefore, to learn from the president of the American Cartographic Society that the circulation of maps of the Middle East increased ten-fold in the United States after the wholesale movement of American troops into the region (National Broadcasting Company, 1990). The Americans were not alone in mapmaking endeavors: Shortly after the Iraqis invaded Kuwait in 1990, their Ministry of Information reportedly published a flurry of Middle East maps (Colvin, 1990, p. 32).

Such overt linkages of maps to power require little denaturalization and, therefore, are of only passing concern in this chapter. Of more interest here is the fact that the Iraqi maps feature Kuwait as the 19th province of Iraq, as it was before the British split it off after World War I (Colvin, 1990, p. 32). For the implicit agenda, the "real uses of most maps," as Wood and Fels (1986) point out, are "to possess and to claim, to legitimate and to name" (p. 72) or, as in the case of the Iraqi maps, to repossess and to reclaim. Giddens (1979) would agree: For him, the legitimation of dominant interests is one of the primary modes in which ideology operates and, therefore, one of the chief concerns of the student of ideology (p. 188). The conclusion of Helgerson's (1986) study of the mapping practices of English Renaissance cartographers reflects this perspective: "The choice they made, the choice of what to study and describe, was, however little sense they may have had of its broader implications, a choice of one system of authority, one source of legitimacy, over another" (p. 58). Such choices, moreover, are not confined to inclusions that legitimate or empower the dominant interests; they may also entail inclusions that *disempower* "the Other." In her study of 17th-century Dutch mapping practice, Alpers (1983) notes the claim by Dutch cartographers Braun and Hogenberg that human "figures were included in their city views to prevent the Turks—whose

religion forbade them to use an image with human figures—from using them for their own military ends" (p. 133).

Rules of inclusion determine, moreover, not only what phenomena are mapped but also which aspects of the included phenomena are represented. Barthes (1970) notes, for example, that the *Hachette World Guide,* dubbed the *"Guide bleu"* by the French, "hardly knows the existence of scenery except under the guise of the picturesque, . . . found any time the ground is uneven" (p. 74). In Barthes's deconstruction, this inclusionary rule is interpreted as the "bourgeois promoting of the mountains," as a manifestation of the desire to legitimate "Helvetico-Protestant morality"—a morality linked to difficulty and solitude (p. 74). Consider, too, the issue of inclusionary rules informing the city feature map. The map must isolate key iconic features to symbolize the city, but which ones? A cursory survey of such maps reveals a focus on icons representing cultural, recreational, civic, and commercial sites. For graphic designer Trieb (1980), the rules of inclusion are clear: The feature map focuses only on what is "positive and desirable"; it manifests an "optimist world view" in its attempt to create "a 'booster' image so positive and intriguing that it will draw visitors" (p. 19). One strongly suspects that the bourgeois morality legitimated by the feature map is not the Helvetico-Protestantism legitimated by the *Guide bleu—la difficulté vaincue en solitude*—but rather the more self-indulgent morality of capitalist consumerism. Note, however, that the fundamental ideological mechanism—the legitimation of hegemonic interests—remains the same, although the particular Marxist mediating code linking the formal elements of a map and its "social ground" may differ (Jameson's term, 1981, p. 39).

Rules of inclusion also determine the strategies and formal devices used to symbolize the aspects of phenomena chosen for representation. Of particular interest here are the representational strategies used to legitimate dominant interests. The primary legitimating strategy is based on the hierarchization of space, for space is not perceived isotropically, i.e., as everywhere having equal value (Arnheim, 1974, p. 30). Because space is perceived anisotropically, the placement of visual elements becomes a way of imparting privilege. Positioning to privilege may be effected in various ways. Privileging through *centering* has not escaped notice in the cartographic literature. Thus Mercator's 15th-century mapping system, still a standard, placed the designer's homeland, Germany, at the center of the map—a Eurocentric strategy quickly adopted across the Continent. Harley's (1988a) generalization of this phenomenon is worth citing in full:

> A universal feature of early world maps, for example, is the way they have been persistently centred on the "navel of the world," as this has been perceived by different societies. This "omphalos syndrome" [Edgerton, 1987, p. 26], where a people believe themselves to be divinely appointed to the centre of the universe, can be traced in maps widely separated in time and space, such as those from Mesopotamia with Babylon at its centre, maps of the Chinese universe centred on China, Greek maps centred on Delphi, Islamic maps centred on Mecca, and those Christian world maps in which Jerusalem is placed as the "true" centre of the world. (p. 290)

Privileging can also be effected through placement on *top*, as seen in the hierarchical representation of organizational structure in charts, whereby administratively higher echelons of personnel are located above the rest. According to Mumby (1988), "the concept of hierarchy has been reified to the point where the structuring of organizations is perceived as 'naturally' occurring from the top down" (p. 88). In fact, given the degree of naturalization of this particular mode of positional enhancement, our readers may be surprised to learn that Australian children often place their continent in the top half of their world maps, i.e., the maps are upside-down for residents of the Northern Hemisphere (Simon, 1987, p. 15).

Finally, privileging is also effected in a series through *ordering,* where the first, and to a lesser extent the last, elements gain distinction. Wood (1987) writes, for example, of the effect of map ordering in *Goode's World Atlas* when nation-states are given "the ontogenetically privileged position of coming first—out in front of a long sequence of maps of the physical environment" (p. 31). Wood obligingly appraises the consequences of this privileging in terms of *naturalization:* "It [privileged positioning] imposes the impression that these nation-states have the same developmental status as landforms and climate, as though the nation-states were just as natural and hence *not* implicatable in any different way from the rains and the winds in the fate of man. It naturalizes the state" (p. 31).

Representational hierarchies, however, need not originate in spatial anisotropy; they may also be traced to other bases of visual saliency—a saliency associated with the legitimation of hegemonic interests. Marin (1988) speaks eloquently to this point in his analysis of a 17th-century map of Paris developed by the king's engineer, Jacques Gamboust:

> The knowledge and science of representation, to demonstrate the truth that its subject declares plainly, flow nonetheless in a social and political hierarchy. The proofs of its "theoretical" truth had to be given, they are the recognizable signs; but the economy of these signs in their disposition on the cartographic plane no longer obeys the rules of the order of geometry and reason but, rather, the norms and values of the order of social and religious tradition. Only the churches and important mansions benefit from natural signs and from the visible rapport they maintain with what they represent. Townhouses and private homes, precisely because they are private and not public, will have the right only to the general and common representation of an arbitrary and institutional sign, the poorest, the most elementary (but maybe, by virtue of this, principal) of geometric elements: the point identically reproduced in bulk. (p. 173)

Here, the saliency observed derives from the fact that bigger things (two-dimensional signs) are more impressive than smaller things (point signs), iconic structures more impressive than noniconic ones, different symbols more impressive than repeated identical ones (Gombrich, 1979, pp. 151-156).

Marin's (1988) analysis suggests, moreover, a second source of rules of inclusion, a source based not, as just discussed, on the norms and values of "the order of social and religious tradition," but rather on those of what he terms "the order of geometry and reason" (p. 173). For rules of inclusion based on the objective order of science may also be viewed as serving sectional—in this case, disciplinary—interests. Wood and Fels (1986) denaturalize the inclusion of conventional scientific visual forms and devices on the map: "It's not just pragmatism or objectivity that dresses the topographic map with reliability diagrams and magnetic error diagrams and multiple referencing grids, or the thematic map with the trappings of F-scaled symbols and psychometrically divided greys. It's the urge to claim the map as a scientific instrument and accrue to it all the mute credibility and faith that this demands" (p. 99). Furthermore, let us focus on just one representational device mentioned by Wood and Fels, namely, the grid. In doing so, we turn our attention, in terms of Gestalt psychology, from the *figure,* already revealed as complicitous, to the *ground,* which may initially seem especially innocent. In fact, to the extent that the background of the map seems particularly innocent, it is an even worthier candidate for denaturalization—and, ultimately, for viewing as complicit with mechanisms linked to power and authority. In the case of the map, the culprits are the rules governing the transformation of a three-dimensional object

Figure 3.1. Sketch Illustrating the Distortions of the Mercator Map
SOURCE: Kaiser (1987, p. 13).

(e.g., the world) into a two-dimensional representation (e.g., a map of the world). The primary transformational device is the projection scheme chosen—a formal device that inevitably introduces distortion and, as Harley (1988a) notes, "can magnify the political import of an image even when no conscious distortion is intended" (pp. 289-290). Reconsider, in this context, the case of the Mercator world map. The Eurocentric Mercator map manifests clear distortions: For example, Russia, with 22 million km^2 of area, looks twice the size of Africa, with 30 million km^2 (Figure 3.1). For Peters (1983), the most vocal critic of the Mercator map, such distortions serve the sectional interests of the white colonialist powers (p. 63).

In any event, whatever projection scheme is chosen for a mapping, the transformation is embodied in the cartographic space as a *grid,* e.g., as lines of equal latitude and longitude, and this too is complicit with mechanisms linked to power and authority. The overlay of this representational device on the mapping surface has been associated by researchers with diverse ideological notions. Edgerton (1987) observes, for example, that by the time of the Renaissance the Ptolemaic cartographic grid had become a talismanic symbol of Christian authority "expressing nothing less than the will of the Almighty to bring all human beings to the worship of Christ under European cultural domination" (p. 12). Similarly, Konvitz (1990) notes the significance of the regularization imposed by the grid in 18th-century French maps:

> Just as national geodetic surveys bearing a minimum of local references lent themselves to an interpretation of space as a fluid medium for the diffusion of power, urban maps which described a street network at the expense of property lines and the architectonic volume of buildings lent themselves to an interpretation of space which emphasized movement and the economic potential of development and growth. (p. 11).

Rules of Exclusion and Repression

Clearly, our discussion of rules of inclusion serving to legitimate dominant interests demands one of rules of exclusion and repression, for "ideology not only expresses but also represses" (J. Thompson, 1984, p. 85). An inclusionary rule such as "centering of the dominant interest," for example, entails the marginalization of nondominant interests. For some social critics, the study of exclusions rather than inclusions is, in fact, paramount. As Belsey (1980) notes:

> Ideology obscures the real conditions of existence by presenting partial truths. It is a set of omissions, gaps rather than lies, smoothing over contradictions, appearing to provide answers to questions which in reality it evades, and masquerading as coherence in the interests of the social relations generated by and necessary to the reproduction of the existing mode of production. (pp. 57-58)

We turn our focus to that which is excluded, that which is repressed. Once again, we begin with explicit rules of exclusion in mapping practices, because "the map that is not made . . . warrants as much attention as the map that is made" (Monmonier, 1982, p. 99). Indeed, if the very decision to map is fraught with ideological implications, as we have already noted, the same can also be said of the decision *not* to map. In Wood and Fels's (1986) formulation, if "to map a state is to assert its territorial expression, to leave it off [is] to deny its existence" (p. 64). While unmade maps have understandably received scant attention in the cartographic literature, restrictions on the publication and dissemination of maps have been widely recognized. Harley (1988b) discusses numerous cases: In Renaissance Portugal, for example, pilots faced the

death penalty for giving or selling charts to foreigners (p. 61). Harley
(1988a) also provides provocative examples of exclusionary practices
at the level of cartographic aspects, including the omission of the locations of nuclear waste dumps from official topographical maps of the
United States Geological Survey (p. 289; Monmonier, 1991, pp. 118-122).

Synchronic Perspective

As with the explicit inclusionary rules, explicit exclusionary rules
require little denaturalization and, similarly, receive only passing attention here. We turn, then, to implicit exclusionary rules. Prime targets
for exclusion are members of the nonhegemonic groups, what is commonly referred to in the critical literature as the Other. In organizational
terms, this practice is exemplified in the exclusion of the lower echelons
of personnel from organization charts; in class terms, it is exemplified
in the exclusion of the lower class from maps. In the latter instance, as
Harley (1988b) observes in an analysis of 16th- and 17th-century
mapping practices,

> social taxonomy seems to have underlain the silence in European cartog-
> raphy about the majority social class. For map makers, their patrons, and
> their readers, the underclass did not exist and had no geography, still less
> was it composed of individuals. . . . The peasantry, the landless labourers,
> or the urban poor had no place in the social hierarchy and, equally, as a
> cartographically disenfranchised group, they had no right to representation
> on the map. (p. 68)

Others consider the exclusion of what may be termed *the otherness of the
Other*. Trieb (1990), for example, writes of the exclusion of the *living conditions* of the Other: "Feature maps . . . do not show us what is undesir-
able. We never see slums, buildings in poor condition, suggestions of danger.
The feature map is an optimistic world view, an image which focuses on
only the positive aspects of urban, and in some instances, rural life"
(p. 19). And Marin (1984), of the exclusion of the *production practices*
of the Other: "All the productive and transforming activities have been
concealed. The city [17th-century Paris] is not to be read as a workplace. It
is, rather, the place of virtue and glory, commerce and exchange" (p. 214).

Moreover, strategies of containment may involve not only exclusions
but also repressions (Jameson, 1981, p. 213), and frequently what are
repressed are individual differences. Differences can be repressed, for

Figure 3.2. French Map of the Channel Region That Employs Exonyms

instance, through the strategy of *typing*. Thus in the *Guide bleu*, "men exist only as 'types' "—a strategy Barthes (1970) decries as a "disease of thinking in essences, which is at the bottom of every bourgeois mythology of man (which is why we come across it so often)" (p. 75). One is reminded, in this connection, of the observation by Marxist critic Williams (1960) that "there are in fact no masses; there are only ways of seeing people as masses" (p. 300).[1]

Differences can also be repressed by naming practices, for naming—the use of a proper name in particular—is "the ultimate signifier of difference, of uniqueness" (Williamson, 1978, p. 51). Earlier, we viewed the labeling of Kuwait as the 19th province of Iraq as an inclusionary act in the service of legitimation of the dominant interest; here we view it obversely as an act of repression of the otherness of the Other, for to label Kuwait as Iraqi Province 19 is to place it in a homogeneous series of Iraqi territories, thereby denying its national autonomy. Consider, too, mapping practices associated with the use of exonyms, that is, "the rendition of geographic names in a language not having official stature in the region in question and which differs from the local rendition" (Ormeling, 1980, p. 332; authors' translation). Thus, for example, international maps designed in the past by French cartographers often designated Germany with the French word *Allemagne* rather than the German *Deutchland* (Figure 3.2). The supplantation of indigenous forms with foreign versions amounts to a repression, by linguistic appropriation, of the otherness of the Other. Recognizing this, the United Nations is committed to suppressing the practice of exonyms on international maps (Ormeling, 1980, p. 333).

Repression of the otherness of the Other also occurs when that Other is situated in a space ethnocentrically conceived to have homogeneous qualities. Consider that one of the first initiatives of the French Revolution "was to devise a rational system of administration through a highly rational and egalitarian division of the French national space into 'departments' " (Harvey, 1989, p. 255; Figure 3.3). Above all, this reform sought in administrative boundaries "the best model to abolish privilege and eliminate precedent" and thereby to achieve "what had eluded an absolute monarchy: the uniform application of law and administration throughout the nation" (Konvitz, 1990, p. 5). However, such changes are never arbitrary, natural, or innocent; rather, they imply a new structure of power (Raffestin, 1980, p. 153).

Diachronic Perspective

Strategies of repression operate, moreover, on more than the synchronic level considered so far. They can also operate on the diachronic level because, although the map as a concrete graphic text is an act of enunciation with ideological dimensions, such an act takes place in a social context and the map is thus also both an act of production and an act of reception. The map, in other words, may be considered as *process* rather than *product,* and strategies of repression take the form of the repression of process in map discourse.

Repression of the Act of Production. Consider first the map as act of production: In this case, strategies of repression can take the form of repression "in some stricter Hegelian sense of the persistence of the older repressed content beneath the later formalized surface" (Jameson, 1981, pp. 213-214). De Certeau (1984) analyzes the progressive repression of the act of production in the map:

> In the course of the period marked by the birth of modern scientific discourse (i.e., from the fifteenth to the seventeenth century) the map has slowly disengaged itself from the itineraries that were the condition of its possibilities. The first medieval maps included only the rectilinear marking out of itineraries (performative indications chiefly concerning pilgrimages), along with the stops one was to make (cities which one was to pass through, spend the night in, pray at, etc.) and distances calculated in hours or in days,

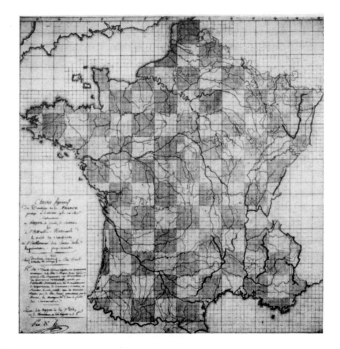

Figure 3.3. Map Used by the National Assembly in Planning the Administration of the First French Republic
SOURCE: Archives Nationales de France, Paris.

that is, in terms of the time it would take to cover them on foot. Each of these maps was a memorandum prescribing actions. . . . Between the fifteenth and the seventeenth centuries, the map became more autonomous. . . . it colonizes space; it eliminates little by little the pictural figurations of the practices that produce it. (pp. 120-121)

Production practices are now inscribed only in highly sedimented, or vestigial, form.[2] De Certeau speaks of the presence of *narrative* figurations (ships, animals, human figures) that function to indicate the operations underlying the making of the given map. Thus, for example, "the sailing ship painted on the sea indicates the maritime expedition that made it possible to represent the coastlines" (de Certeau, 1984, p. 222). And while de Certeau would prefer a fuller representation of production practices on the map, he clearly appreciates the retention of vestigial forms. However, the presence of even these vestiges of the production act has engendered

criticism from other quarters. Information designer Tufte (1990), for example, who shows little appreciation for the ideological dimension of
visual discourse, seemingly would remove all such traces of production
practices:

> Symbols of Scheiner's patron and religious order decorate those areas
> without spots in a hundred such diagrams, a reminder of Jonathan Swift's
> indictment of 17th-century cartographers who substituted embellishment
> for data:
>
>> *With savage pictures fill their gaps,*
>> *And o'er unhabitable downs,*
>> *Place elephants for want of towns*
>
> These symbols, similar to a modern trademark or logotype, may have
> served as a seal of validation for the readers of 1630. Today they appear
> somewhat strident, contradicting nature's rich pattern. (p. 21)

Not surprisingly, the trend toward suppression of the act of production is paralleled in a repression of the *agent* of production. Jacobi and
Schiele (1989) note the suppression of the scientist in scientific journals:

> Scientists are never portrayed in primary scientific journals, and it is out
> of the question to publish photographic portraits. The reason for this is easy
> to understand: science is enunciated without reference to the enunciator.
> The author disappears behind an object that seems to speak for itself, or
> write itself out independently. (p. 750)

The enunciator is equally suppressed in the map. No longer, for example,
do we encounter the patron portrait which figured so prominently in the
cartographic cartouche of yore. Rarely do we even encounter on the map
the name of the designer, much less its prominent placement in the manner
of the Minard maps.

Repression of the Act of Reception. Turning from the map as act of
production to the map as act of reception, we encounter equally repressive strategies at work. If earlier we viewed the map as the totalization
of past itineraries, here we view it as the totalization of all potential
itineraries or narrative journeys; the critical question becomes: How is
the viewer inscribed in such discourse? or, in Althusserian terms, How

is the concrete individual interpellated as subject? In answer, we consider the influential design of the critically acclaimed London Underground Diagram (LUD). There, as in routing diagrams generally, the viewer is interpellated not as a concrete individual with a specific potential itinerary in mind, but rather as a contemplative observer—the master of abstract possibilities—and while she might appear to be empowered through ascription of an Olympian perspective, such empowerment is illusory: As a specific user with concrete travel needs, she is relatively unempowered.

What goals are served by the ascription of such an Olympian perspective? To answer this question, we must consider the representational strategies of the LUD in more detail. These strategies are particularly clear when viewed from a historical perspective. The LUD has as a forebear the underground (subway) map in use in 1924, which faithfully presented geographic relations among the many stations (Figure 3.4). In consequence, the individual lines appear to meander and the portion of the diagram devoted to central London is seriously congested. This map is, moreover, in the nature of a palimpsest in that the origins of the underground as a multiplicity of independent, uncoordinated enterprises and facilities are readily detected. In contrast, a recent LUD, based on a design by engineering draftsman Beck, is ostensibly a highly schematized representation of the system that sacrifices geographic accuracy in the interests of both readability and consonance with experience (Figure 3.5).

Thus the LUD distorts not only by representing the individual lines with graphical elements oriented either vertically, horizontally, or diagonally but also by comparatively outsizing the central region, with the overall impression of a rather homogeneous structure when judged from apparent separations between stations. Katz (1988) rationalizes this homogenization on the basis that "experientially, the distance between stops is the same: the rapid transit experience is measured in stops not time or distance. . . . The scale adjustment, or shift, in the [LUD] is 10 times—an inch in the country equals ten times the distance in the city" (p. 2). Moreover, in another homogenization, the use of standard symbols to represent stations masks the considerable physical differences among them.

Many critics have lauded the resulting clarity and intelligibility of the LUD (Arnheim, 1974, p. 159; Garland, 1969, p. 79; Katz, 1988, p. 2; Walker, 1979, p. 2). But the clarity is deceptive; it is attained, as we have shown, at the cost of considerable distortion. More important here,

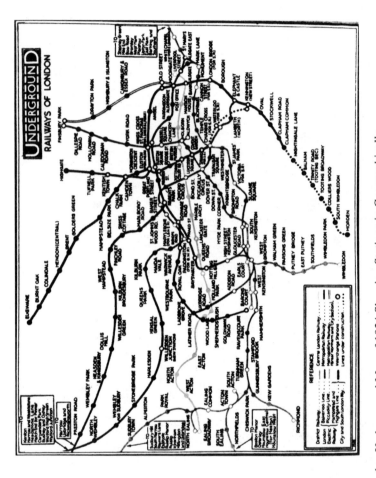

Figure 3.4. London Underground Map of 1924 Showing Stations in Geographically Correct Relations

SOURCE: London Transport.

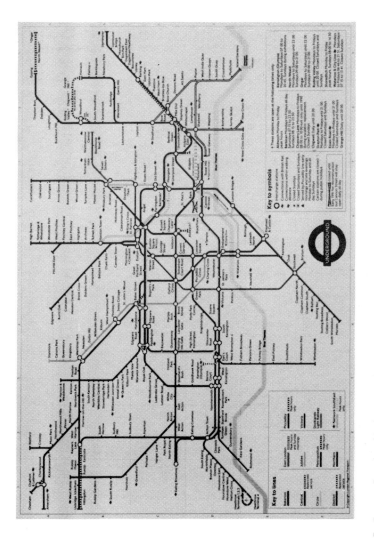

Figure 3.5. Recent London Underground Diagram

SOURCE: London Transport.

however, it is attained at the cost of considerable dissimulation. Dissimulation of what? And dissimulation to what end? In effect, Trieb (1980) answers the first question: "Due to the powerful effect of the underground map, many visitors sense the structure of urban London as the structure of the underground. . . . The diagrammatic structure of the underground map is clear, articulate, and legible—all the things that London as a city is not" (p. 15). For an answer to the second question—dissimulation to what end?—we recall Barthes's (1970) association of visual discourse with the mythological discourses of the bourgeois class, especially those seen to foster nationalism and the fetishization of commodities (p. 127). Certainly Trieb's comment would sustain a reinterpretation in terms of nationalism, i.e., the disparity between the clarity of the London of the LUD and the messiness of the actual city can be interpreted as a dissimulative attempt to engender chauvinism in the viewer. For an association of the LUD with the "fetishization of commodities," we turn to Forty's insightful account relating the act of LUD reception to capitalist consumerism: For Forty (1986), the purposive distortions of the LUD "induced people to undertake journeys they might otherwise have hesitated to make. . . . It is impossible to say to what extent London Transport's design policies excited people's appetite for travel, but there seems every reason to believe that by making travel seem easy, effortless and enjoyable, they contributed to the very substantial leisure traffic" (pp. 237-238). In short, claims Forty, the viewer of the LUD is confronted with an "object of desire," not an "object of use." And that viewer, we might add, is correspondingly interpellated as consumer rather than as user.

Denaturalizing Mapping Practices

Certainly the deconstructive cast of this chapter so far has served to reveal the complicity of map discourse with ideology, but simply to reveal the radical entanglement of ideology and maps may not be enough. Recall Peters's (1983) deconstruction of the Mercator map as an expression of the superiority of Caucasians, an interpretation shared generally by the cartographic establishment (p. 63). Unfortunately, as cartographer Robinson (1985) observes, the Mercator map is "still regularly displayed. Millions see it daily, since it forms the backdrop for the ABC and NBC evening television news programs, as well as others, and it is embarrassing to see it used similarly in the briefing rooms of both the Department of State and the Pentagon, which appear frequently on TV news presentations" (p. 109). Clearly, denaturalization of the Mercator map by the cartographic professionals has not sufficed. Rhetorical theorist Bizzell (1990) points out the limits of a purely deconstructive mode of analysis:

> In their deconstructive mode, the anti-foundationalist critics do point out the effect of historical circumstances on notions of the true and good which their opponents claim are outside time. In other words, the critics show that these notions consist in ideologies. But once the ideological interest has been pointed out, the anti-foundationalists throw up their hands. And because they have no positive program, the anti-foundationalist critics may end up tacitly supporting the political and cultural status quo. (p. 667)

We are thus prompted to go beyond establishing the ideological interest underlying mapping practice to projecting a positive program for new map design methodologies. What would such a program entail? Barthes's (1985) indictment of naturalized sign systems provides a hint: "There is an evil, a social and ideological disorder, ingrained in sign systems which do not frankly proclaim themselves as sign systems" (p. 66). One solution, then, is to design maps that *do* frankly proclaim themselves as sign systems. What would be the *desiderata* in such maps? What, in other words, are some of the denaturalizing mapping techniques that might imbue a new design practice?

To some extent, the call is for a broader palette of mapped texts. In this spirit, Harley (1989) calls for a "new cartography" that would replace "the arid academic cartography"—a new cartography characterized by "a greater pluralism of cartographic expression" to "serve the breadth of social and historical problems that now engage our attention" (p. 88). But more is not necessarily different. Consider, for example, the proffered alternatives to the Mercator world map. Peters (1983) offers an alternative map based on an equal-area projection scheme, a map that, in his words, "is an expression of the basic equality of all peoples in this post-colonial era" (p. 145), a map that he also unfortunately claims "is entirely free of any ideology" (p. 148). The United Nations has, in fact, adopted Peters's *Third World Map* for many purposes. However, according to Robinson (1985), the Peters map eliminates one set of biases only to introduce others. Robinson's alternative, a compromise between geographic fidelity and aesthetics, has in its turn been the target of similar charges. And so it goes.

In short, what is really needed is a new politics of design, one authorizing heterodoxy—a politics where difference is not excluded or repressed, as before, but valorized. This ideal of encompassing contestatory as well as hegemonic discourses has been variously expressed in the literature of postmodernist theory. Barthes (1986) calls for the inclusion of "acratic" discourse (discourse outside power, which repudiates the doxa) as well as "encratic" discourse (discourse within power, which conforms to the doxa) (p. 120); Bourdieu (1977), for the construction of "extraordinary discourse" or counter discourses (p. 170); Venturi (1966), for the privileging of "elements which are hybrid rather than 'pure,' " "both-and" over "either-or", for "messy vitality over obvious unity" (p. 22); Foucault (1969/1972), for a view of discourse as a "space of multiple dissensions," a space characterized by provisionality and heterogeneity rather than totalization and homogeneity (p. 155). Harvey (1989) sees such concern for *alterity,* or otherness, as the "most liberative and therefore the most appealing aspect of postmodern thought" (p. 47).

Such inclusion of the Other will necessitate, in turn, a new vision of the map. After all, when Huck Finn's notion of the map as reality, or even that of the map as mirror, proved a disabling perspective, we were led in our deconstuction of the map to adopt the more enabling metaphor of map as text, more specifically, as "textually homogeneous site of ideological inscription." Now, concerned with the proposal of denaturalizing practices, we suggest two new metaphors: For addressing the suppression of the spatial or synchronic perspective, we propose the metaphor of the *map as collage;* for addressing the suppression of the temporal or diachronic perspective, we propose the metaphor of the *map as palimpsest.*

Synchronic Perspective: The Map as Collage

The denaturalizing power of the collage—an assemblage of diverse elements drawn from preexisting texts and integrated into a new creation manifesting ruptures of various sorts (Group *Mu,* 1978, pp. 13-14) —has not escaped notice in the theoretical and practical postmodernist discourses (Derrida, 1979; Hassan, 1987, p. 445; Kuspit, 1989, p. 46; Ulmer, 1983, p. 88). In an analysis of postmodern photography, Hutcheon (1989) describes the collage effect induced by the invasion of words into the semantic space traditionally dedicated to the pictorial and the consequent denaturalization:

> Postmodern photo-graphy [*sic*] works to "de-doxify" by making both the visual and the verbal into overt sites of signifying activity and communication. It also contests the glossing over of the contradictions that make representations (linguistic or pictorial) serve ideology by seeming harmonious, ordered, universal. Its paradoxes of complicity and critique, of use and abuse of both verbal and visual conventions, point to contradiction and, thereby, to the possible workings of ideology. (pp. 138-139)

Unlike the photograph, visual genres such as the graph and the map are to some extent already collages, of course, because they are traditionally composed of both alphanumeric and figural signs, but we are advocating an even stronger melange of sign systems. A strong commitment to the collage technique is, incidentally, not without historical precedence in cartography; it was favored by Minard, a 19th-century French engineer and graphics innovator, whose demographic maps often bore tables, references, notes, and text (Figure 3.6). An admirer of Minard's work, Tufte has exploited the collage technique to great effect in the design of his landmark work on information display (Barton & Barton, 1990). A theoretical commitment to the collage effect on various levels is also evident in his book: On the level of the individual graphic, he encourages us to "write little messages on the plotting field to explain the data, to label outliers and interesting data points, to write equations and sometimes tables on the graphic itself, and to integrate the caption and legend into the design" (Tufte, 1983, p. 180). On the level of the page, Tufte (1983) advocates a breakdown of the semantic spaces dedicated to text or graphic: "Tables and graphics should be run into the text whenever possible, avoiding the clumsy and diverting segregation of 'See Fig. 2,' (figures all too often located on the back of the adjacent page)" (p. 181).

Nor is the collage effect confined to unexpected juxtapositions of linguistic and pictorial systems; juxtapositions may also occur *within* either the linguistic or pictorial systems. Russian formalist Uspensky (1973) points out the defamiliarization effect induced by the juxtaposition of different linguistic systems within a given textual space (pp. 20-32), e.g., by the insertion of French in (the Russian) *War and Peace,* whereby language loses its ostensible transparency as a universal signifying system and is thereby seen as a culturally relative phenomenon. The United Nations' policy of suppressing exonyms in favor of indigenous place

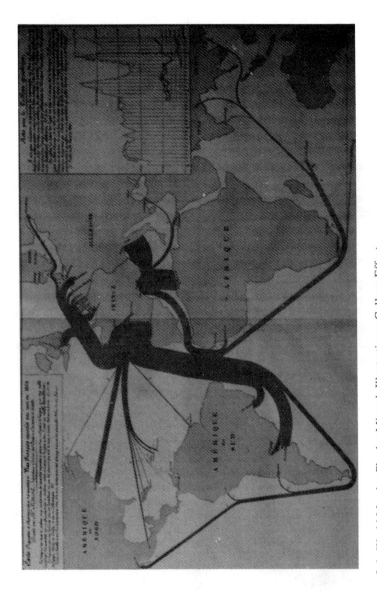

Figure 3.6. World Map by Charles Minard, Illustrating a Collage Effect

SOURCE: Collection de Documentation et de Communication, L'École Nationale des Ponts et Chaussées, Paris.

names on world maps and the attendant juxtaposition of differing points of view—the hegemonic and the contestatory—is clearly relevant here.[3] As an example of the juxtaposition of pictorial systems, we cite Bryson's (1983) illustration of the way "paintings can violate the existing doxa by representing within the same frame discursive practices thought to be disparate" (p. 42; Shapiro, 1988, p. 150). Within the mapping tradition, one finds relevant examples in the indigenous scenes inserted long ago in cartographic space. Moreover, our taxonomy of collage strategies is far from complete, for it is not our intent here to present a comprehensive taxonomy, much less an analysis, of the myriad ways in which heterogeneity of representational orders can create a collage effect. Lynch and Woolgar (1988) suggest the scope of such an endeavor by mentioning the heterogeneities resulting from the juxtaposition of various representational devices, theoretical principles, or representational functions, e.g., resemblance, symbolic reference, similitude, abstraction, exemplification, or expression (p. 100).

Diachronic Perspective:
The Map as Palimpsest

Reculons pour mieux sauter. Thus far in our discussion of denaturalization techniques for the map, we have advocated a paradigmatic shift to a heterogeneity of representational order induced by the juxtaposition of various representational systems—a juxtaposition metaphorically captured by the expression *map as collage*. Indeed, construing the map as collage proved a useful heuristic for discovering a radical design practice that would juxtapose communication elements laterally in seemingly anarchistic fields and thereby accord representational status to hitherto excluded or repressed interests. Because it is a synchronic, or spatial, notion, however, the collage is ultimately limited, for, as we noted earlier, strategies of repression also operate in diachronic, or temporal, perspectives where corresponding counterstrategies of denaturalization are equally needed. In his discussion of the representation of place on maps, de Certeau (1984) recognizes the limited heuristic value of the collage metaphor: "The place, on its surface, seems to be a collage. In reality, in its depth it is ubiquitous. A piling up of heterogeneous places. Each one . . . refers to a different mode of territorial unity, of socioeconomic distribution, of political conflicts and of identifying symbolism" (p. 201). "The kind of difference that defines every place is not on the order of a juxtaposition but rather takes the form of imbricated strata" (p. 200). In short, de Certeau concludes, "the place is a palimpsest" (p. 202). The map as palimpsest metaphor enables us to address the kinds of repression that occur in a diachronic perspective, i.e., the repression of the acts of production and reception of the mapped text considered vertically rather than laterally, considered as process rather than as product, as speech act rather than as structure.

Denaturalizing the Act of Production

How does one denaturalize the act of production? For Harkin (1989), it means making "one's discursive strands visible, like the plumbing and ductwork of a postmodern building" (p. 56). In less figural terms, it means resisting strategies designed to reify one's text by allowing a franker representation of the layers in which the text is historically and inevitably imbricated. In the field of document design, denaturalizing the act of production means challenging many tenets of current publication policy. It may mean, for example, resisting the increasingly popular policy of suppressing the set-off, extended quotation, which visually acknowledges the alterity of the Other, in favor of the seamless web of text produced by the paraphrase, which more fully but less overtly appropriates the Other. It may also mean resisting the equally popular publication policy of suppressing the footnotes and marginalia that contextualize one's work and that visually establish the ongoing relation of one's thought with past thought. Tufte's practice of placing references and notes alongside the relevant text material is laudable in this respect and, given its ease of implementation in this era of desktop publishing, is certainly available to all. Harley's (1989) praise of the *Historical Atlas of Canada* serves as a testimonial to the value of the palimpsest technique in mapping practice:

> Part of the success of the acclaimed *Historical Atlas of Canada* [R. C. Harris, 1987] is that the Editors have broken the link with the narrow dogma of academic cartography. Bar graphs, flow lines and proportional circles survive but they are enriched by architectural and archeological drawings, by original maps and town views of the past, and by landscapes with people and artefacts. We begin to know what it might have felt like to have lived in old Canada. A narrative unfolds expanding rather than denying the rhetorical power of the map. Even the most austere maps can become a *pictura loquens* as other images help to trigger their meaning. (p. 88)

Denaturalizing the act of production also means not closing off the movement of contradictions by representing meaning as fixed and stable. The design tradition, of course, has been to *stabilize* meanings, but the designers of visuals can just as easily choose to *destabilize* meanings. In the field of medical illustration, cardiologist Wahr (1987) advocates the inclusion in journal articles of the *multiple* radiographic views taken during a cardiac catheterization procedure to assess coronary artery disease, not just the *one* best supporting a given research thesis. For a cartographic example, we return to the issue of representing Kuwait. According to Gilbert Grosvenor, president of the National Geographic Society, if the status of Kuwait remains in question when the first update to the sixth *Atlas of the World* is printed, the regional map will include a statement of who occupies the territory and the United Nations' position in the situation (cited in Christian, 1990). For Wood (1986), destabilizing meaning may entail more than the addition of an explanatory note to a map: Rejecting the conventional distinction between reference and narrative atlases, he praises the latter for the fact that in "committing themselves explicitly and narratively to a point of view, [they] become thereby full (not empty); and, in admitting what they were about, become, *through this gesture,* truthful in a way 'reference atlases'—hiding their messages behind the false front of 'objectivity'—never have" (p. 38). Wood presumably would favor the incorporation in reference atlases of the running text characteristic of narrative atlases, thereby denaturalizing the ostensible objectivity of the former's point of view.

Denaturalizing the Act of Reception

How, in turn, does one denaturalize the act of reception—the other occluded layer of the map considered as process? How, in other words, does one denaturalize the traditional view of visuals as objective artifacts that speak for themselves? Clearly, one can begin by more frankly acknowledging that looking is an interpretive act. Once again, Minard's maps offer an exemplary strategy: They bear notes telling the viewer how to interpret the map. Recall, also, Tufte's (1983) similar advice to "write little messages on the plotting field to explain the data" in quantitative graphs (p. 180). More generally, denaturalizing the act of reception means producing positions that enable subjects to occupy realistic rather than imaginary relations within the social totality. Earlier in our discussion of the LUD, we noted a relative disempowerment of the

viewer as a specific user with concrete travel needs—a disempowerment masked by the illusory power accrued through ascription of an
Olympian perspective inscribed in the map. The viewer's relative disempowerment becomes immediately obvious when one considers an
alternative, more enabling representation—the lighted board map in
several Parisian Metro stations whereby the viewer can override the
totalizing effect of the map by registering her destination and receiving
an individualized, highlighted itinerary.[4] "You are here" indications on
maps provide a second, admittedly less powerful, example.

Denaturalizing the act of reception also means adopting the perspective of the traditionally disempowered. Earlier we noted the exclusion
of lower echelons of personnel from organization charts; here we advocate not simply their inclusion but, rather, the adoption of their point
of view. Technical-communication educators Mathes and Stevenson
(1976) propose, for example, replacing the conventional organization
chart with an "egocentric organization chart" (Figure 3.7) as a way of
enhancing the position of, or empowering, the writer: "The egocentric
chart categorizes people in terms of their proximity to the report writer
rather than in terms of their hierarchical relationship to the report writer.
Readers are not identified as organizationally superior, inferior, or
equal to the writer, but rather as near or distant from the writer" (p. 15).
Reporting on the field of cartographic education, Minerbrook (1991)
notes that the Texas Education Agency has recently required a comparison of different map projections in new textbooks for secondary schools
(p. 60). Moreover, the Peters projection, favoring the Third World point
of view, is now routinely included in the map presentations of Paramus,
New Jersey, schools.

Conclusion

Our proposal of a more inclusionary visual design practice may con-
jure up visions of an ocular cacophony of juxtaposed and superimposed
elements, devices, and practices. Clearly, the governing aesthetic of the
visual as collage-palimpsest is not the modernist "less is more" but
rather the postmodernist "less is a bore"—an aesthetic that privileges
complexity over simplicity and eclecticism over homogeneity, an aesthetic that tends toward the fragmentary and the local, an aesthetic that
renounces the driving ambition toward Unity with a capital "U" and

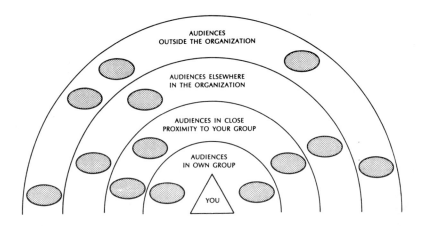

Figure 3.7. Egocentric Organization Chart
SOURCE: Reprinted with the permission of Macmillan Publishing Company from *Designing Technical Reports* by J. C. Mathes and Dwight W. Stevenson. Copyright © 1976 by Macmillan Publishing Company.

"disperses itself among discreet claims and observations" (Bryson, 1989, p. 705). Unity may still be achieved, but it will be a hard-won unity—a unity eschewing the reassuring grand synthesis in favor of the uneasy collocation of competing and rival claims, the difficult unity of inclusion rather than the easy unity of exclusion—perhaps, in fact, the only meaningful unity for our time (Barton & Barton, 1987).

Notes

1. On the other hand, this is not to say that the alternative is to view people as obviously unique, autonomous, and totally self-determining individuals, for that would be to succumb to what Althusser (1971) has termed "the elementary ideological effect" (p. 161). Rather, in Althusser's famous phrase, "ideology interpellates concrete individuals as subjects" (p. 160).

2. Similarly, Bazerman (1988) notes a trend toward increased sedimentation of experimental physics procedures in journal articles. In a study of visuals in articles published early and late in the career of *Physics Review,* he notes the progressive exclusion of visuals more closely reflective of laboratory practice. His documentation of the decreasing use of "detailed apparatus drawings and extensive tables of raw experimental data" in favor of "extensive equations and schematized graphs" clearly reveals a trend toward suppression of the production act in scientific visuals (p. 172).

3. Note that the suppression of exonyms is neither new nor necessarily easy. According to Ormeling (1980), "As early as the end of the nineteenth century, cartographers working on an international series of world maps . . . decided to adopt exclusively the naming system peculiar to each country. But there is a world of difference between good intentions and execution. Whether one likes it or not, each national cultural politic commits itself, in the name of the mother country, to preserving its own exonyms rather than sacrificing them to facilitate communication" (p. 333; authors' translation).

4. M. Burke and McLaren (1981) report the recent introduction of electronic displays by the London Transport at certain mainline rail stations. The displays "present a menu of approximately three hundred routes which may be taken from that point. The dialogue with the system is in three languages simultaneously (English, French, and German). The selected route is presented in isolation from the total network; points where the passenger is obliged to change train are stated clearly" (p. 109). Moreover, as M. Burke and McLaren note, "theoretically the system could be extended by being distributed to domestic television sets by means of a viewdata or teletext system" (p. 109).

INTERPRETATION

4

Formalism, Social Construction, and the Problem of Interpretive Authority

THOMAS KENT

Following the lead of philosophers and historians like Nietzsche, Dewey, Heidegger, Wittgenstein, and especially Kuhn, professional writing teachers and writing theorists tend generally to accept nowadays the claim that knowledge is socially determined. Instead of talking exclusively about knowledge in terms of cognitive processes like brain function (Walker, 1990), logic (Young, Becker, & Pike, 1970), recall (H. Clark & Haviland, 1977), or thinking procedures (Flower & Hayes, 1981), writing teachers more and more frequently talk about knowledge in cultural and historical terms and fall in line with rhetoricians like Bruffee who claim that knowledge derives from the socially constructed—and, therefore, historically determined—discourse communities in which we live. Because we no longer believe that knowledge is something that exists only in our heads, we are led also to say that the production of discourse, especially in settings outside the academy, comes into being through social interaction, because discourse can be recognized as discourse only after it becomes part and parcel of the normative conventions that form the social communities in which we all must live and work.

One way to view the significance of this shift from a concern with mental processes to a concern with social and historical ones is to see the shift as a change in our conception of interpretive authority. Whenever discourse is produced, some aspect of the communicative interchange always will be seen to *authorize* a correct interpretation of an utterance, and depending on our conception of communicative interaction, the agent that authorizes interpretation will change. For example, those who believe that knowledge is something stored in the head, instead of something shared within a community, usually claim that interpretive authority derives either from the text itself or from the mental processes we all employ and hold in common. North (1987, p. 245) calls these people "formalists," and for convenience, I will stick with his term. For my purposes here, two groups of formalists can, generally speaking, be distinguished: *textualists*—who are out of fashion—and *cognitivists*—who are going out of fashion. In *Modern Rhetoric,* Brooks and Warren (1970) demonstrate clearly the presuppositions about interpretive authority held by textualists:

> There are four basic natural needs that are fulfilled in discourse. We want to explain or inform about something. We want to convince somebody. We want to tell what a thing looked like—or sounded like, or felt like. We want to tell what happened. These natural needs determine the four forms of discourse. Each need represents, then, an intention that is fulfilled in a particular kind of discourse. (p. 56)

Brooks and Warren's now largely discredited theory of modes or "forms of discourse" employs a fairly straightforward Kantian conception of discourse production. According to Brooks and Warren (1970), we all share certain mental categories that can be expressed as modes of discourse, so effective communication amounts to finding the correct mode to convey our intentions. To know the text, therefore, means that we know the writer's intention or mind. In this formulation, a direct and unproblematic connection exists between the text and the writer's intention. Meaning, therefore, always lies within the text because the text stands as the representation of the writer's intention. When communication breaks down—when a text, let us say a set of instructions, is misinterpreted—the textualist will almost always diagnose the problem as the writer's failure to find the correct mode in which to express the writer's intention. If a set of instructions proves faulty, the problem lies within the text, and the writer will need to tinker with the text until the writer's intention becomes clear to a reader. For the textualist, the text always authorizes interpretation in the sense that the text is the signifying element mediating between writer and reader and finally the only place where conflicting interpretations may be adjudicated.

The cognitivist, on the other hand, shifts the seat of interpretive authority from the static text to the dynamic mental processes that supposedly generate the text. Cognitivists claim that texts derive their comprehensibility from certain codifiable cognitive activities that all competent writers and readers hold in common, so texts, in a sense, become the representations of the cognitive processes employed by writers and readers. North's (1987) account of the assumptions held by cognitivists is worth noting here:

> All the Formalist work in Composition I can find has treated writing as a cognitive process: the operation of the mind conceived as a system working to gather information from outside of itself, sift, store and retrieve it in some way, sort and arrange it [to] meet various criteria, and eventually transform it into words on a page that accomplish some set of goals. (p. 245)

Although North addresses the emphasis that cognitivists place on the mental processes of writers, we should remember that cognitivists also investigate the mental processes of readers. For example, in a well-known article concerning revision strategies, Flower, Hayes, and Swarts (1983) demonstrate perfectly the presuppositions about interpretive authority held by most cognitivists when they investigate what they take to be the mental processes employed by readers that enable them to comprehend a functional document. Flower et al. (1983) tell us that "this article describes a research project intended to uncover some of the things readers need in order to process, or comprehend, a functional document" (p. 42). After Flower and co-workers "uncover some of these things readers need," they draw a specific conclusion about the reading process: "Our study of reading protocols revealed a revising principle used by readers: they try to structure information around agents performing actions in specific situations" (p. 49). From this conclusion, they move on to offer suggestions for writers of functional documents, suggestions such as "Organize around action, not terms," "Organize around the reader's questions," and "Use headings with a human focus" (p. 54). In their account of these "reader-based revisions," the presuppositions employed by Flower et al. (1983, p. 52) about the nature of meaning jump out at us. Like all cognitivists, they presuppose that (1) meaning constitutes a process that occurs in the head—in this case, the head of a reader, (2) these cognitive processes may be isolated and codified, and (3) from an account of these processes, researchers can draw conclusions about the nature of writing and reading that, in turn, may be employed to draw up guidelines for good writing. For cognitivists, a breakdown in interpretation means that something has gone wrong with the writer's understanding of either the reader's or the writer's own mental processes. For example, if a reader fails to comprehend adequately the steps in a set of instructions, the writer cannot correct the fault, as the textualist would claim, simply by tinkering with the text. Because interpretive authority resides in the cognitive processes we employ in order to understand the text and not in the text itself, such a breakdown in communication, according to the cognitivist, can be corrected only by addressing the thinking processes employed by the writer or the reader.

On the surface, the social constructionist view of interpretative authority would seem to constitute a radical break with formalism. On the one hand, formalists—both textualists and cognitivists—claim that a determinate, stable, and ahistorical component of communicative interaction exists—a component such as a mode of discourse or a codifiable cognitive process of one kind or another—to which we can refer to help settle conflicts in interpretation. On the other hand, social constructionists claim that no such transcendental element exists; according to the social constructionist, interpretive authority always resides within the historically determined discourse communities in which we live. From this perspective, professional writing instruction—our systematic accounts of the nature of effective communicative interaction—is determined by the changing writing practices that take place in what are perceived to be specific discourse communities. Bruffee (1982) tells us, for example, that—from the social constructionist point of view—teachers

> see our job as helping our students identify their own beliefs and the beliefs of the communities they belong to; helping students examine how knowledge communities express and justify beliefs; helping students discover communities of belief relevant to their interests; and helping students learn to join, to maintain, and, when appropriate, to move from one knowledge community to another, or to disband a knowledge community altogether. (p. 111)

Because writing always occurs within the practices of historically determined discourse communities that change from time to time and from place to place, the interpretations we give texts will also change from time to time and from place to place. Consequently, according to the social constructionist, nothing transcendental—nothing ahistorical —exists to which we can appeal to settle disagreements about the correct interpretation of a text. When communication breaks down, the fault lies exclusively neither in the head nor in the text; the fault lies in our misunderstanding of the changing social conventions that structure the different communities in which we live.

In summary, the social constructionist objection to formalism concerns the formalist contention that an agent that authorizes interpretation exists outside of the normative conventions that structure the different discourse communities where communication occurs. Stated a bit differently, social constructionists object to the essentialism inherent in formalism; they object to the notion that writing possesses an essence that can be described as a codifiable textual strategy such as a theory of modes or as a cognitive process. According to social constructionist thinking, when formalists reduce writing to an ahistorical systemic process, they cannot account for certain crucial aspects of discourse production—aspects such as context, gender, class, and institutional authority. Consequently, formalists can never relate a complete story about the nature of discourse production; their work is always incomplete. Breaking with the idea that something ahistorical exists—a cognitive process or a textual strategy—to which we can refer to settle our disputes about meaning, social constructionists believe that they can provide a more complete and coherent account of discourse production than formalists can provide.

In the discussion to follow, I want to suggest that this social constructionist view of professional writing is not so different from the formalism it attacks so that, finally, the social constructionist explanation of interpretive authority actually encounters the same difficulties as the formalist explanation. Because my critique of formalist and social constructionist conceptions of interpretive authority depends heavily on Davidson's philosophy of language, I will begin with a short account of his attack on the idea of conceptual schemes, and then I will move on to suggest an alternative to formalist and social constructionist conceptions of interpretive authority, an alternative that I call paralogic hermeneutics.

Formalism, Social Construction,
and the Problem of Conceptual Schemes

Although social constructionists claim to break new ground in the study of discourse production by attempting to account for the social dimension of interpretive authority, I believe that social constructionists actually perpetuate the ideas that they wish to overturn. By claiming that interpretive authority derives from the discourse communities in which we live, social constructionists endorse the idea of a conceptual scheme. As Davidson (1986a) explains, conceptual schemes are "ways of organizing experience; they are systems of categories that give form to the data of sensation; they are points of view from which individuals, cultures, or periods survey the passing scene" (p. 183). For the social constructionist, interpretation always will be relative to the conventional framework or *conceptual scheme* that holds together a specific discourse community. As a result of this relativism, a conceptual scheme of one kind or another always will be seen to mediate between us and the world; a text can mean only what our conceptual schemes allow the text to mean. Although these conceptual schemes change from time to time and from place to place, we nonetheless find ourselves at any *particular* moment prisoners of some sort of conceptual scheme. We find ourselves cut off from a world "out there" beyond our conceptual schemes, for all we can ever know are the conceptual schemes that mediate between us and the world.

Conceived broadly, the social constructionist conception of interpretive authority and the formalist conception share a common philosophical position: Both directly endorse the Cartesian claim that a split exists between a comprehensible realm composed of conceptual schemes— modes of thought, cognitive processes, or social conventions—and an incomprehensible realm composed of things-in-themselves that exist outside of our conceptual schemes. For the formalist, interpretive authority derives from the modes of discourse or cognitive processes we all share, and these processes, in turn, constitute the frameworks or windowpanes through which we survey the passing scene, a passing scene that we may know only by looking through some sort of mental windowpane. Consequently, faulty interpretations occur when writers or readers employ inappropriate modes of discourse or incommensurate cognitive processes; authoritative or correct interpretations occur when writers and readers employ the same modes of discourse or cognitive processes or, we might say, when they share the same mental framework or cognitive windowpane. For the social constructionist, interpretive authority is established in practically the same way. According to social constructionists, interpretive authority derives from the social conventions that form the frameworks for the different discourse communities in which we all live. In this formulation, faulty interpretations occur when writers and readers employ the wrong social conventions; authoritative interpretations occur when writers and readers employ the social conventions that fit the particular interpretive frameworks employed by the disparate discourse communities in which they live. Finally, the only consequential difference between the formalist and the social constructionist formulations of interpretive authority concerns the location of the authorizing agent. For the formalist, the authorizing agent resides in the head; for the social constructionist, the authorizing agent resides in historically determined social conventions. Both, however, accept the Cartesian claim that something—either mental states or social conventions— mediate between us and the world.

From the perspective outlined above, both formalism and social construction endorse the fundamental assumption that a conceptual scheme or framework of one kind or another allows writing to come into being. For both the formalist and the social constructionist, writing only becomes comprehensible or capable of interpretation within a framework of either shared mental states or shared social conventions, and this shared framework exists anterior to the act of writing. To communicate effectively, writers must learn to employ these frameworks or they will not be able to produce discourse that others can comprehend. In this scenario, writing—as a kind of communicative interaction —becomes a derivative systemic process that either imitates a cognitive activity or reflects a set of social norms. In other words, writing— according to both formalists and social constructionists—may be reduced to specific codifiable processes that model either cognitive activities or conventional social behaviors, and these processes, in turn, show up in our textbooks and form the foundation for our professional writing courses.

Several serious problems exist, I believe, with these formalist and social constructionist conceptions of interpretive authority, conceptions that follow directly from the presupposition that the act of writing is relative to a conceptual scheme. Here, I cannot outline all of these problems, but I would like to consider what I take to be the most serious difficulty—the claim that interpretive authority derives from a framework

HEORY, AND RESEARCH

58   ...ng act. (For further discussion of these is-
         1989b, 1991.) To spell out as clearly and suc-
         ...c the problem with these framework theories, I will
         ...ncated version of Davidson's critique of the conceptual
    ...argument along with a general account of his theory of commu-
nicative interaction.

Beyond Formalism and Social
Construction

Challenging formalists and social constructionists who maintain that
interpretive interaction derives from some sort of conceptual scheme—
modes of discourse, cognitive processes, or social norms—Davidson
(1986a) argues that conceptual schemes derive from our interpretive
interaction and not the other way around as formalists and social
constructionists would have it. In his attack on the claim that all we may
ever know is a subjective realm of mental states or a framework of
socially constructed conventions, Davidson insists that communicative
interaction places us in direct contact with other minds and with a world
of objects and events so that nothing—no conceptual scheme or framework—mediates between us and the world. Although Davidson's argument for this position is complex, I believe that—for my purposes here
—we need to concentrate only on his primary objection to the conceptual scheme argument. As I have noted already, conceptual relativists—
such as formalists and social constructionists—argue that meaning,
knowledge, interpretive authority, and so forth are relative to a conceptual scheme, so moving from one conceptual scheme to another means
that we must learn new meanings, confront new knowledge, and encounter different authorizing agents of interpretation.

Davidson argues, however, that this position is contradictory. Referring to the man who is generally credited with inventing social construc-
tionism, Davidson (1986a) explains that "[Thomas] Kuhn is brilliant at
saying what things were like before the revolution using—what else?—
our post-revolutionary idiom" (p. 184). Davidson's point here is decep-
tively simple: Conceptual schemes cannot be incommensurate with
one another; if they were incommensurate then someone like Kuhn
could never comprehend with any certainty the writings of Aristotle or
Newton. This same critique clearly applies to the formalist argument:

If meaning, knowledge, and interpretive authority are commensurate with internal mental states like modes of discourse or cognitive processes, then how may we be confident—as we always are in day-to-day life—that we know either the minds of others or the world of objects and events outside our own subjectivity? The fact that we do know well enough the minds of others—others who talk as we do and others, like Aristotle and Newton, who don't—contradicts the formalist's and social constructionist's fundamental claim that we can know only the conceptual scheme in which we find ourselves and, consequently, that we may never know directly the minds of others or objects in the world. (For a discussion of Davidson's other objections to conceptual relativism, see Dasenbrock, 1991, and Rorty, 1989.) For my purposes, the upshot of Davidson's argument is that formalists and social constructionists cannot claim, on the one hand, that interpretation is relative to a conceptual scheme and claim on the other hand, that these interpretations tell us anything about the minds of others or objects in the world, which, of course, our interpretations clearly do. Even more to the point, the formalist and the social constructionist cannot explain, without begging all the important questions, how it is that we can get things done in the world through written communication—for instance, following a set of written instructions—if all we can ever know is a conceptual scheme.

Davidson, of course, denies that interpretation is relative to a conceptual scheme or to a conventional framework of social norms that we must learn before communication can occur. Davidson (1986d) argues that "what interpreter and speaker share, to the extent that communication succeeds, is not learned and so is not a language governed by rules or conventions known to speaker and interpreter in advance" (p. 445). According to Davidson, the element of communicative interaction that allows us to get things done in the world is something he calls a "passing theory" or what I have called "hermeneutic guessing" (Kent, 1989b). By *passing theory,* Davidson means the tentative guess or theory we employ either to understand an utterance or to produce an utterance. He tells us that "the passing theory is [what] the interpreter actually uses to interpret an utterance, and it is the theory the speaker intends the interpreter to use" (Davidson, 1986d, p. 442). Rorty (1989) explains that "a theory is 'passing' because it must constantly be corrected to allow for mumbles, stumbles, malapropisms, metaphors, tics, seizures, psychotic symptoms, egregious stupidity, strokes of genius, and the like" (p. 14). Because a passing theory "must be constantly corrected," it cannot be predicted in advance of a communicative situation, for we often must adjust a passing theory on the spot as we communicate. As a result of the contingent nature of interpretation, there can be no rules for generating a passing theory, so the act of interpreting an utterance cannot be reduced to a system or to a framework that we can know in advance of a communicative situation. Davidson (1986d) explains that

> linguistic ability is the ability to converge on a passing theory from time to time—this is what I have suggested, and I have no better proposal. But if we do say this, then we should realize that we have abandoned not only the ordinary notion of a language, but we have erased the boundary between knowing a language and knowing our way around in the world. For there are no rules for arriving at passing theories, no rules in any strict sense, as opposed to rough maxims and methodological generalities. A passing theory really is like a theory at least in this, that it is derived by wit, luck, and wisdom from a private vocabulary and grammar, knowledge of the ways people get their point across, and rules of thumb for figuring out what deviations from the dictionary are most likely. There is no more chance of regularizing, or teaching, this process than there is of regularizing or teaching the process of creating new theories to cope with new data in any field—for that is what this process involves. (pp. 445-446)

In place of conceptual schemes, the authorizing agent of interpretation for Davidson (1986d) becomes the tenuous passing theory that cannot be constructed from "rules in any strict sense, as opposed to rough maxims and methodological generalities" (p. 446).

From a Davidsonian perspective, nothing codifiable exists—no framework, no scheme, no process, or no epistemology—that will ensure a correct interpretation of our utterances. All we possess are our tentative passing theories or hermeneutic guesses—the interpretive activity that I have called "paralogic hermeneutics" (Kent, 1986b)—that allow us to relate our meaningful marks and noises to the meaningful marks and noises of others. In the case of professional writing, for example, our knowledge of different frameworks or processes never will ensure in advance of a communicative situation that a text—a set of instructions, for instance—will be interpreted adequately. We may know the framework of generic conventions typically employed to construct instructions, we may know how to arrange these conventions according to rules of information processing, we may know the contextual framework for the instructions and the readers who will use the instructions, we even may field test the instructions before we submit them to the intended readers. However, our knowledge of these frameworks or processes cannot ensure—in advance of a particular communicative situation—that the instructions will work. On the other hand, to say that a passing theory cannot be reduced to a framework or codifiable process certainly does not mean that background knowledge is unimportant, for clearly a correlation exists between our background knowledge and our ability to produce an adequate passing theory. When we share a great deal of background knowledge with one another—or when "people tend to speak much as their neighbors do," as Davidson (1986b, p. 278) puts it—we are much more likely to generate passing theories that our neighbors will be able to interpret efficaciously. Although our background knowledge clearly helps us generate effective passing theories, we should remember that background knowledge—the knowledge of generic conventions, cognitive processes, social communities as well as the knowledge of more fundamental elements of communication such as spelling, syntax, and paragraph construction—should not be confused with the *employment* of this background knowledge. To communicate by *employing* our background knowledge, we generate passing theories that may be modified when we discover that misinterpretation has occurred. From this paralogic hermeneutic perspective, writing—as a kind of communicative interaction—does not correspond to the different kinds of background knowledge that we teach in our classrooms. Instead, writing is conceived as a thoroughly hermeneutic activity that cannot be reduced to a process or framework and then taught.

When we locate interpretive authority within the paralogic give-and-take of communicative interaction and not within conceptual schemes such as cognitive processes or systems of social conventions, the problem of misinterpretation takes on a new light. From a paralogic hermeneutic perspective, faulty interpretation occurs because communicants cannot converge on an adequate passing theory and not because communicants lack a particular kind of background knowledge. For example, an ineffective instruction manual might or, then again, might not be caused by the writer's ignorance of certain cognitive processes or certain social conventions. A writer's understanding of a cognitive process such as top-down processing or a writer's understanding of the context of social conventions in which a manual will be used does not warrant a correct interpretation. A writer may understand every guideline ever written in every professional-writing textbook and still manage to produce an ineffective instruction manual. This annoying fact exists because passing theories never match precisely, for if they did match precisely, misinterpretation would be an impossibility. We should remember, however, that most of the passing theories we employ in day-to-day life match so closely with our neighbors' passing theories that everyday communication, in a sense, becomes transparent. Until we run into a communicative situation that requires conscious interpretation, we hardly realize that we are interpreting one another's utterances at all.

Although I have been arguing here that professional writing—as a kind of communicative interaction—cannot be taught, especially in the way we currently imagine that we teach it, I remain convinced that our traditional professional writing courses nonetheless serve a useful function. Unlike many academic writing courses, professional writing courses generally provide students the opportunity to test their individual passing theories while helping students, at the same time, to develop new background knowledge that they may find helpful in different communicative situations. Of course, I also believe that professional writing teachers will provide even more helpful guidance to their students when they no longer authorize interpretation through appeals to cognitive processes or frameworks of social conventions. For those of us who regularly teach professional writing, this shift from talking about writing as either a process or a conventional act to talking about writing as a hermeneutic interaction would challenge us, I believe, to drop our current process-oriented vocabulary and begin talking about language-in-use, language that helps us get things done in the world. For the professional writing classroom, collaboration might replace teacher-centered instruction, and no longer would the teacher occupy center stage. The professional writing teacher would become an adviser or, better yet, a consultant and would hold no privileged body of knowledge that the student would need to ferret out, because no privileged body of knowledge—no privileged framework or process—exists. Instead, the teacher would consult actively with the student to help the student generate effective passing theories and, in turn, effective written texts.

The view of interpretive authority that I am advocating here—and the professional writing pedagogy arising from it—obviously would pose marked problems both for our institutions and for our discipline. For our institutions, traditional professional writing courses would be eliminated, and in place of our current process-oriented instruction, teachers would advise students on an individual basis. Collaborating with the student, the professional-writing teacher would be thrown into specific communicative situations, and the teacher and the student together would engage in communicative interaction with others within and outside the university. In addition to the changes in the traditional professional writing classroom, a shift to this paralogic-hermeneutic instructional method also would create complex problems for the discipline of professional writing. For example, such a shift would require a wholesale change in the way we currently think about professional writing pedagogy. We would need to acknowledge that writing does not take place solely in the head or in discourse communities and that writing cannot be reduced to a systemic process. Ultimately, we would need to accept the unsettling claim that writing *instruction* is a misnomer, for no body of knowledge in the area of writing instruction exists to be taught. In addition, we would need to make the difficult admission that we cannot instruct students to become good writers because *good writing*—as a transcendental category—does not exist. When we begin to see writing as a paralogic act—a kind of communicative interaction that cannot be reduced to a systemic process—*good* writing coincides solely with our ability to arrive at an accurate hermeneutic guess. Good writers are good guessers, and the interpretive authority we employ to recognize a good guess resides solely in the efficacy of our guesses. As I indicated previously, we certainly can improve the efficacy of our guesses by knowing something about different modes of discourse, cognitive processes, and normative social conventions, but finally, interpretive authority does not reside within any of these conceptual schemes. Instead, interpretive authority resides in our ability to use language to get things done in the world, and we get things done in the world through the paralogic give and take of human communicative interaction.

INTERPRETATION

5

Generic Constraints and
Expressive Motives:
Rhetorical Perspectives on
Textual Dialogues

JOSEPH J. COMPRONE

This chapter brings together evolving theory and practice from two broadly defined and distinct perspectives on learning to write: contextual linguistics and genre theory as fostered by the contextual or Hallidayean linguistics now being given a good deal of attention in the United Kingdom and Australia (Halliday & Hasan, 1985) and expressionistic rhetoric as it has been developed by Britton and his followers in the United Kingdom, the United States, and Australia (Dixon, 1987, pp. 11-12).[1]

My purpose in bringing these research areas together is to provide a foundation for complex treatments of genre in technical writing courses. In many such courses, genres are treated as templates within which rhetorical processes are to fit, rather than as socially constructed forms that are themselves in a continual state of reconstruction and revision. This chapter will open up the concepts of genre to a dialogue focused on the questions of when and how we, as technical writing teachers, can make considerations of genre part of the composing processes of our students. I also argue that rhetorical analysis of context must become a more integral part of our technical writing courses.

Some background theory must first be established. This theory is best provided through a review of recent work on dialogic literacy, which, better than any other single theoretical field, serves as a base on which expanded notions of genre can stand.

Literacy:
The Expressive and Social Perspectives

Underlying this contrast between contextual linguistic and expressionistic perspectives is a much broader professional conversation concerning definitions of literacy—of how literacy works and why it works as it does. In school research in many countries (the United States, Australia, the United Kingdom, among others), social definitions of literacy have come to oppose expressive definitions. Social theories, often developed out of work by theorists such as Langer (1987), Vygotsky (1962), and Foucault (1966/1971, 1969/1972, 1975), emphasize the interdependency of text and context, of subject and object, and of the socially constructed and individual aspects of composing. Expressive theories focus primarily on the writer as source and manager of textual expression, with emphasis resting squarely on the choices writers make as they work against the grain of social convention (Bartholomae, 1985, pp. 152-153). This polarity has become even more extreme in composition studies in the United States, where recent social constructionist work has replaced the expressionistic at the center of teaching and research programs.[2]

What has happened in recent years to place these social-constructive and expressionist contraries into a new theoretical context? First, literacy studies—and the teaching, family, and workplace practices evolving from them—have made it impossible for informed theorists to separate reading, writing, and speaking as they consider individual composing processes or institutionalized text-making practices (Cooper & Holzman, 1989, p. x; Heath, 1986, p. 217; Langer, 1987, p. 3; Szwed, 1981, p. 23). Second, work in literary theory, particularly the work in reader response and gender criticism, has made it difficult for critics or teachers of literature to isolate writing and reading processes; those

processes are now almost always considered as interdependent parts of
a larger social process (Bakhtin, 1981, p. 254; Bleich, 1988, p. 330;
Fish, 1980, p. 327). Third, and finally, because of speech act theory,
teachers and critics are no longer able to treat literature as a special or
elite form of discourse (Riffaterre, 1978, pp. 86-87), at least not without
meeting with a good deal of argument from literary researchers and
theorists alike (Fish, 1980, p. 97). Speech act theory, more than any
other approach, has enabled theorists to understand literature as defined
by its intended social function rather than by a particular author's
"higher intention" (M. Pratt, 1977). Both Fish and Pratt argue that
meaning in literary works involves as much negotiation of generic
forms and interpretation of intent as any type of "ordinary discourse."

All three fields support a move away from theories of literacy that
oversimplify the conventions behind a field of discourse and instead
encourage a more dynamic sense of generic forms. These various developments support a move away from dichotomous theories of literacy;
in particular, they strongly suggest that the debate in the 1970s and early
1980s between social and expressive theories was reductive and that
the models used by today's teachers, researchers, and theorists of literacy should emphasize as far as possible the interdependency of text and
context and the need to balance perspectives on context (social and
cultural) and individual language users (students and professionals) as
we develop programs and do research.[3]

These three areas of research—literacy studies, literary theory, and
speech act theory—have combined to provide a base for new, less dichotomous, more integrative perspectives on literacy. They can show
teachers of technical and professional writing how to give appropriate
attention to individual cognitive and expressive processes without un-
dervaluing the important social function of genre. This balance is particularly important in technical writing courses where the generic forms
and organizational contexts must constantly influence one another. (See
Mathes and Stevenson's, 1976/1991, textbook for a good example of
the complex array of organizational contexts facing technical writers
and the potential effects of these contexts on conventional technical
writing genres.) Teachers will be able to develop workable strategies
for integrating convention and intention only when they have a clear
idea of when and how genres and individual intentions interact within
composing processes. Within the dialogic framework, these two perspectives—the social and the individual—overlap. The context within
which the writer composes shapes the patterns and forms through which

individual texts develop and function (Bizzell, 1986a, p. 66). Within this larger framework, individual texts are influenced by several contextual factors: by the purposes of broader discourse communities; by audiences, or particular collections of people brought together by either professional or social interests; and by function, or the particular use to which a text will be put by the people who read it. But individual composers are the ones who must control and manage these social processes every time they compose.

Work by the Russian postformalists Vygotsky and Bakhtin provides a sound base on which to construct dialogic theories of literacy. Vygotsky (1962, pp. 9-24) argues against Piaget's genetically based, developmental theories of language development. Children, Vygotsky (1962) argues, possess the capacity for cognitive development toward abstraction, but that capacity serves only as a mental framework within which actual language learning occurs (pp. 33-41). That learning itself, Vygotsky contends, actually occurs only when the child verbally interacts with trusted adults; in other words, the potential for language development is activated only when social interaction occurs (pp. 44-51). Vygotsky proposes the notion of "inner speech" (p. 44) to explain how a child's internal, cognitive processes are first, in infancy, monologic (the child literally talks to herself in a silent, inner voice) and then gradually become dialogic by incorporating the other, usually parents or teachers, into that inner monologue. After a while, both inner and outer voices have merged into a genetically based but socially constructed inner dialogue.

Where do "adult" conventions such as genres come in here? In their responses to children, adults often place a child's personal language into the frames and schemes familiar to the adult language-user. Such responses help get the work of the world done, and they help the child function effectively. But, of course, in every child-adult interaction social negotiation occurs. The generic form represented in the adult response is put in negotiation with the child's intention. The process of talking through activities as basic as shoe tying becomes a means of learning how to balance the pressures of genre and intention, of learning how to respond in complex ways to situations that involve both conventional and individual response (Vygotsky, 1962, pp. 136-137). Individual and social consciousness are inextricably integrated.

Bakhtin, in his work on the novel and on speech genres, carries the implications of Vygotsky's theories of early cognitive development into the more general area of adult higher literacy. In *The Dialogic*

Imagination, Bakhtin (1981) analyzes centuries of narrative development within the novel genre. He argues, using a post-Sausserian linguistic framework, that this genre, from classical folk narratives through the
early modern, has always served as a repository for what Bakhtin calls
"heteroglossic text." Such texts are marked by the bringing together of
different linguistic registers and fields, by a mixing of different social-
political and ideological languages in a text unified by themes that cross
these sociolinguistic boundaries. In fact, Bakhtin valorizes the novel
precisely because it transcends specific generic boundaries in order to
render a more global image of the social macrocosm.

Bakhtin expands our notions of genre through his work on the novel.
But he attempts to refocus our sense of the complexities involved in
using the concept in his work on speech genres (Bakhtin, 1986). In this
work, Bakhtin gives teachers of communication in all fields a concept
of genre that is both theoretically flexible and rhetorically specific. This
is accomplished through a reexamination of the process through which
sentences are turned into utterances and utterances are defined by the
whole discourses in which they function (Bakhtin, 1986, pp. 29-37).

To summarize, Bakhtin (1986) argues that genres are extremely complex and plastic forms that are constructed through social interaction
(p. 66). Every time an individual produces an utterance within a conversation, that utterance immediately begins to interact with its context.
The contact point, as utterance and context meet, is genre; the conventional form embodied in the genre interacts with the intention embodied
in the individual utterance. In both directions, change occurs. The utterance is shaped into a sentence that is intricately tied in with the context
provided by the whole discourse, which, in turn, is influenced by the
utterance.

This complex interaction is not limited to supposedly more open
forms of discourse. In fact, technical writing may provide the clearest
example of how this balance between intention and genre works. When
writers in organizational settings attempt to carry out some kind of
ordinary discourse, they are even more directly affected by context and
genre. In most corporate situations, there are forms at every turn. These
forms are in constant interaction with the need of individuals or groups
to carry out particular tasks. Often these tasks demand unique or individualized responses, but the generic forms that are conventional to
the contexts in which these writers function are also necessary if the
organization is to continue achieving common goals. The most effective
writers in such situations are able to treat intention and genre as dynamic

counterparts, one dependent on the other, but in often contrastive ways. In addressing multiple audiences, for example, a technical writer will often revise even the most basic of memo forms to provide each audience with a different rhetorical emphasis.

Both Vygotsky and Bakhtin have been linked with social constructionist perspectives on the epistemologies of technical, professional, and literary textual fields (Bazerman, 1988, pp. 291-317; Comprone, 1990, pp. 56-63; Winsor, 1990b, p. 60). It is at this point of conjecture between socially constructed epistemologies, textual heteroglossia, and social perspectives on language learning that important questions concerning genre arise. Addressing these new questions concerning the teaching and criticism of genre could become the foundation for developing approaches to writing and reading that embrace both the individualistic focus of the expressive theorists and the more contextual focus of the social constructivists.

The work of both Vygotsky and Bakhtin shows teachers the way toward dialogic literacy by opening up both cognitive and sociolinguistic perspectives on genre. Genres to Vygotsky are not genetically reinforced schemata that children develop at different rates as they grow. Rather they are socially reinforced speech acts through which individual intentions and social process are mediated. Genres to Bakhtin are not socially defined linguistic patterns that are arbitrarily imposed on individual utterances. Rather they are conventions that serve as points of interaction between individual intentions and the broader contexts within which any writer works.

Genre and Social and Expressive Perspectives

Recent work on genre in Australia and the United Kingdom adds yet another dimension to this developing dialogic theory. Often going by the name of *contextual linguistics,* this movement emphasizes the need for teachers of literacy to focus on the point in composing discourse at which the intentions of writers must engage conventional generic forms. This engagement, in turn, can only be fully understood when contexts are analyzed rhetorically. With this understanding, dialogic theory would suggest, comes the potential to play with genres, to see them for what they are—necessary social agencies in what is essentially a drama that places individual agents and social agencies in creative tension.

What is important for teachers of reading and writing to recognize in this approach to genre is the fact that contextual linguistics strives to describe the creative tension between the formal, linguistic features of a text and the social situations within which they operate, and it places composers into contexts in which they can achieve their purposes by balancing social convention and individual intention. Contextual linguistics (Halliday, 1988, p. 31; Threadgold, Grosz, Kress, & Halliday, 1986) has set the features of particular texts within the larger framework of communities, conventions, and their interactions, and it has placed socially negotiated meaning at the center of the process of textual construction and reconstruction (Halliday, 1975).

Several key concepts describe this linguistic approach. Linguistic concepts of *field,* or the activities shared by members of an institution and the topics used to categorize those activities; *mode,* or the medium through which meanings are communicated—spoken, written, or visual—and *tenor,* or the distance or degree of formality and informality established between writer or speaker and audience (Martin, 1986, pp. 236-240, 243-244) interact within a particular culture's linguistic system to create a *register* (Halliday & Hasan, 1985). Linguistic registers describe the interactions of field, mode, and tenor within a given cultural context, while *genre,* to contextual linguists, represents the different forms, dynamic and always in transition, that texts take as they are influenced in the making and receiving by these different levels (field, mode, and tenor) in a linguistic system. For the contextual linguist, this complex process of managing overlapping linguistic codes and systems is at the center of composing or interpreting texts. Individual composers and interpreters are not working solely in an introspective way, nor are they working in a context solely dominated by social convention. Rather, they are best viewed as textual managers.

Contextual Linguistics and
the Genre Debate in Australia

Opportunities, of course, always create conflict and risk as well as new and constructive approaches. This fact is apparent in work by school language researchers and teachers in Australia, where contextual linguists Martin (1986), Christie (1985), Kress (1985), and Rothery (1984) have been engaged in an ongoing debate with Australian writing researchers such as Dixon (1987, p. 9) and Reid (1987, p. 3), who support an expressive approach to learning language. Contextual linguists who support genre approaches argue that they are simply reestablishing the centrality of Malinkowski's "context of situation" (cited in Christie, 1987, p. 25). They contend that students must learn how to read situations before they can manage to express personal meaning and that, in fact, there are no totally personal expressions, because all expression is largely constrained by generic expectations of discourse communities (Christie, 1987, p. 25).

At the bottom of this disagreement is a distinction between two very different perspectives on choice. Dixon and the expressionists adhere to the idea that choice is always rooted in individual will, that however much rhetorical choices may be constrained by genre and context of culture or situation, individuals can at some critical point in composing a text break from those constraints and make independent choices. The genre theorists and contextual linguists also emphasize rhetorical choice, but they argue that choices grow out of knowledge (often tacit) of the values and conventions of communities (Kress, 1987, p. 35). Composers make original choices only when they are thoroughly steeped in the thinking, feeling, and text-making strategies and conventions of their discourse communities (Christie, 1987, p. 27). In fact, genre theorists in most cases would argue that rhetorical choices are the result of a writer's ability to manipulate linguistic and formal patterns that are passed on from one situation to another (p. 27).

In *The Place of Genre in Learning* (Reid, 1987), Dixon (1987, p. 17) and Sawyer and Watson (1987, p. 51) reinforce the expressive perspective. Although they accept many of the social-constructive premises of their genre-oriented Australian colleagues, Dixon, and Sawyer and Watson in essence criticize the genre perspective for its overemphasis on teaching and prescribing preexisting, socially regulated, and restrictive forms. Dixon (1987) argues that introducing genre through form-oriented assignments in elementary schools too often imposes socially constructed genres on learners before they understand themselves and their intentions well enough to work with these genres in complex and interdependent ways (p. 15). He also argues that, in mature discourse, genres themselves are renegotiated every time someone writes and that the key to effective literacy teaching is to empower students to change as well as learn the genres within which they work. Introducing students too early to genre-driven assignments cuts off their ability to influence the contexts within which they speak, write, and read (Dixon, 1987,

p. 17). Dixon advocates an approach in which he "as a teacher set[s] up or reinforce[s] a range of generic choices" that, in turn, interacts with "joint enquiry by students" (p. 17).

Sawyer and Watson (1987) echo Dixon when they argue that the genre approach too often emphasizes the simple copying of linguistic features of generically defined texts (p. 47). These authors contend that, although the approach of Australian genre theorists emphasizes continual social interaction and change within and among genres, this approach often unwittingly encourages pedagogy that reduces reading to mechanical feature analysis and writing to superficial transfer of linguistic features from one textual situation to another. Rhetorical choices are limited to a narrow range of linguistic features, and purposes and audiences are not seen as contextual variables in complex situations that writers can recreate through rhetorical choice but as static formulas to be applied in any situation. Concepts and meanings, which are the important ends of literacy-learning to expressive theorists, are learned by mastering a knowledge of conventions *and* by combining that mastery with the ability to use dialogue, negotiation, and individual intention to change and adapt those conventions. As Sawyer and Watson (1987) claim, "students acquire a new genre in the performance of it —in the struggle to solve a problem worded in a particular way, in the struggle for meaning—not by being given explicit instruction beforehand" (p. 49).

Both sides agree that to write effectively writers must understand the rhetorical forms and features common to situational contexts, they must know the patterns of form that others writing in similar contexts have used, and they must be able to fit their own intentions into the rhetorical patterns they create as they compose. Individual expression, the genre theorists imply, occurs within this larger social matrix, not as isolated verbal action with the individual pitted against the grain of social convention (Kress, 1987, pp. 43-44). Teachers encouraging the genre approach argue that students must learn the conventions of those communities *before* they can make contributions to them, original or otherwise; teachers advocating the expressive approach begin with individual exploration and move much later to considerations of genre and social context. This emphasis on context thus unites theory and practice by linking social awareness and individual intention in the process of composing.

If contextual linguists are right and all writing is ultimately socially negotiated, then where and how does originality come in, and how do genres themselves change? Kress (1987) speaks for most contextual linguists when he explains that original texts come about when composers belong to different and overlapping discourse communities and are able to manage the codes and forms of these overlapping communities in ways that construct new textual genres (p. 37).[4] Once individual readers and writers are able to absorb and apply the conventions and contexts of different communities, they are also able to transpose the forms and patterns common to one community into the texts they address to other communities, and they are able to accomplish this without destroying the coherence of their texts. They are, in other words, able to create what Bakhtin (1981) has called a heteroglossic text (pp. 259-422) and what I have called a plural text—texts that effectively manage the overlapping patterns and strategies of various discourse communities (Comprone, 1990, pp. 58-59).

Martin, Christie, and Rothery (1987, p. 68) outline the pedagogical approach of genre theorists. This approach begins by laying out a definition of genre: "a staged, goal oriented social process" (Martin et al., 1987, p. 59). This definition then becomes the fulcrum on which swings the activities of writing and reading. Learning to write becomes, from this perspective, a matter of negotiation and interaction (p. 69). Teachers create assignments that immerse students in contexts within which they consult with teachers (as representatives of the discourse/knowledge communities found in the assignment) and peers working on similar projects. Within this context, the focus is on student-to-teacher and student-to-student consultation. Teachers act as representatives of professional communities *and* as interpreters of the individual student's intention. Students share each other's struggles to balance generic constraints and personal meanings (pp. 67-72).

Dialogic Literacy, Genre, Rhetoric, and Technical Writing

This review of the literary and linguistic perspectives on literacy, with their respective emphasis on dialogue and genre, brings us to the central question behind this chapter: How can these two fertile perspectives be combined in both theory and practice, producing an integrated and coherent perspective on literacy, useful to critics as well as to teachers of technical writing? The answer, I argue, can be found in

revitalizing and revising rhetorical theory and practice, and in connecting that theory and practice to a revision of our concepts of genre. It is
in rhetorical studies of texts in context that relationships between
linguistic features and social genres can be clarified, and it is within
this clarification of feature/genre relations that the rhetorical choices
of writers and readers can be explained and taught. When, for example,
a linguistic feature such as parallel structure is turned into the rhetorical
trope of parallelism to emphasize conceptual similarity, we have an
example of a linguistic feature used to support rhetorically a writer's
intention. What we learn from tying together research on rhetoric and
genre can inform and guide the way we teach technical writing.

Genre theorists' emphasis on negotiating context and genre through
consulting a teacher-expert contains the seed of a fully developed
but so far unaddressed argument for the primacy of rhetorical choice
and strategy within any complex discourse situation. In fact, Martin,
Christie, and Rothery (1987) argue that teachers must find ways of
introducing "strategies familiar to children from experience of learning
to write" into their plans for teaching writing (p. 68). This emphasis on
developing strategies to guide choice making in the process of producing written text can become the theoretical basis for introducing rhetorical technique and method into writing courses; it can also become a
way of developing an approach to literacy through which the mediation
of socially constructed genres and individual choice making within particular textual situations can be accomplished.[5] With such an approach
in front of us, choice and strategy become central in the students' composing processes, which helps us as teachers use rhetoric to enrich treatments of genres in technical writing courses.

Green makes an explicit call for just such a reapproachment as the
one between genre and individual expression that I call for here. He
focuses on the need to integrate "the contextualist arguments of the
process position [of Britton, Moffett, Dixon, and others] with the text-
centered arguments of the 'genre' position" (Green, 1987, p. 88). For
Green, although he never mentions the word, rhetoric would enter the
debate between genre and process/expression specifically as a method,
first, for examining text in context and, second, as a means of helping
students, workers in business and industry, and others learn how to
make choices that are not prescribed by immediate exigencies alone,
but informed as well by broader social and cultural frameworks. This
developing understanding among teachers of technical writing can help

them know when to intervene in their students' composing processes to help them adapt genre to organizational situations.

Situation, Strategy, Genre, and Choice: Teaching Writing

In this final section, I accomplish two things. First I use the rhetorical thinking of three theorists—Lloyd Bitzer, Carolyn R. Miller, and Dale Sullivan—to provide a basic definition of rhetoric's function within the process of composing technical writing texts. Second, I recommend a series of strategies for technical writing teachers to use as they help students manage the composing process to assure balance between genre and intention. While developing these theory/practice perspectives on rhetoric and composing, I also carry my analysis of genre a step farther.

In 1968, Bitzer began the process of redefining rhetoric in terms of social constructionist theory. His article "The Rhetorical Situation" argued that rhetoricians up to the late 1960s had been working with an impoverished understanding of context, one that emphasized methods employed by authors and orators to meet the needs of immediate or identifiable audiences over strategies and processes for encompassing larger, socially mediated situations (Bitzer, 1968, p. 2). Bitzer argues that rhetorical texts are those that are called into being by situations, that are driven by the social pressures, functions, and needs presented by context: "rhetoric is a mode of altering reality, not by the direct application of energy to objects, but by the creation of discourse which changes reality through the mediation of thought and action" (p. 4).

This redefinition of rhetoric as focused on situation leads directly to an understanding of text making as essentially the dual act of comprehending the context of situation and developing strategies for managing textual production. And textual production is, in turn, a means of acting on situations, a way of changing social contexts by becoming an agent of the discourse represented by those situations (Bitzer, 1968, p. 5). Bitzer's main contribution comes in his use of the term *rhetorical exigence,* which he defines as the issue underlying and motivating the situation within which a text evolves (pp. 5-7). Rhetorical exigence causes a group of speakers, readers, and writers (let us here call them *rhetors*) to function as a social network and to produce texts that address each other within a consistent, systematic framework. But exigence also enables change because within the constraints of the situational discourse driven by rhetorical exigence are embedded the openings, ambiguities, and tensions that individual writers or groups of writers can choose to address in their own texts.

Effective writers must be able to define and understand the rhetorical exigencies that underlie the situations within which they write; they must also be able to find within the discourses produced by those situations those places where arguments can be made, where new information or different perspectives need to be brought to bear. Understanding exigencies means being able to know and participate in the social process within which a particular community of discourses produces texts.

At this point, the concept of genre becomes important. If genres, as current social constructionists argue, are "staged, goal-oriented social processes," then genres are in turn what we might call *discourse management strategies.* Genres are combinations of socially constructed forms—preexisting formats and templates, in many cases—and individually produced responses to those preexisting forms. This is where, however, current discussions of genre would depart from Bitzer's thinking on exigence and situation.

Bitzer locates change within situations that define texts. C. R. Miller, however, who represents more recent thinking on the function of genres, leaves more room for individual discoursers to affect change by their own management of the text-making process. Miller (1984) would agree with Bitzer that genre describes a social process based on the discourser's ability to read the exigence defining the situation within which a text will be presented (p. 151). But Miller goes on to argue that genres as social processes are constantly subjected to the pressures put on them by individual composers. This pressure makes a genre into "a complex of formal and substantive features that create a particular effect in a given situation" (C. R. Miller, 1984, p. 153), and for Miller, this total generic effect can be effected as much by the individual choices of composers as it can by preexisting forms (p. 156). Miller then argues that "genre is a rhetorical means for mediating private intentions and social exigence; it motivates by connecting the private with the public, the singular with the recurrent" (p. 163).

What does this redefinition of genre as social strategy in the act of composing texts imply for the teaching of technical writing? The following three teaching strategies would apply were we to adopt this definition.

First, teachers of technical writing would have to avoid the extreme responses to genres that have characterized their past classroom treatments. Genres are not, as technical writing course materials and textbooks have often in the past suggested, ready-made templates for the mechanical construction of texts (Souther, 1989, p. 4). Nor are they completely open-ended patterns of discourse always at the mercy of ingenious and imaginative manipulators of texts. Genres, as C. R. Miller (1984, p. 156) and D. Sullivan (1990, p. 377) assert, are complex and dynamic processes by and through which writers manage the social conventions and individual intentions they must integrate as they produce any complex text. It is helpful here to point to Sullivan's (1990) argument that genre is not a set of rhetorical skills; it is more accurately conceived of as describing a "practice or praxis" through which composers manage the conventions and intentions they wish to represent in a text (p. 382). This revised notion of genre should encourage teachers of technical writing to have students consider the organizational situations within which they compose.

Second, if genres are best conceived of as discourse strategies, then we as teachers must provide opportunities for students to use them as strategies as they compose. This means including in the classroom composing process opportunities to analyze the contexts that lie behind any rhetorical situation to identify the exigency behind that situation. And it means as well that writers must apply this understanding of exigency to the particular array of social conventions and individual intentions that present themselves within particular rhetorical situations. Genres, from this perspective, are socially constructed forms that must be revised and reapplied in every technical writing situation.

Third, we as teachers must look to our understanding of the complexities of genre as we encourage students to look at every act of writing as a difficult and complex process of balancing and managing convention and intention. Writing processes in most cases can be effectively represented as containing the need to (1) analyze and interpret a rhetorical situation, (2) apply the results of that analysis to produce a set of strategies for managing all the constraints and motives involved in that situation, and (3) use those strategies to make rhetorical choices during the text-making process. This means as well that teachers must develop interactive models for which conversation, oral presentation, reading, individual planning and contemplation, collaborative efforts, and writing come together. Each medium, in other words, brings pressure to bear on whatever genres are conventional to the discourse situation. The result of this interaction of media, as far as composing is concerned, is often a renegotiation of genres to fit them to the needs of individual writers. Theorists of technical writing have supported this rhetorical renegotiation of genre through the interaction of seeing, speaking, and writing (Fearing & Sparrow, 1989).

It is useful to think of writing not as entirely socially or individually motivated, but as a mode of discourse particularly suited to learning how to manage information, ideas, conventions, and intentions. Only with effective management techniques can individuals use writing to find and place their voices in the ongoing conversations that are generated by rhetorical situations; only with effective management techniques can individuals use writing to change the direction of these conversations. The concept of genre can be an effective means of approaching the problem of developing strategies for directing this management process. This balanced and dynamic approach to management through genre is especially important because, as D. Sullivan (1990) points out, "the very thought processes embodied in most modern technical genres have grown out of the technical mindset" (p. 379).

If we as teachers provide students with mechanistic models of genres, we will equip them only to follow what that mind-set has produced in the past. Yet, if we ignore what certain technical situations have produced in the past, we risk producing students who are not able to find a role or place in significant discussion of technical issues.

The answer to this problem can be briefly illustrated with reference to a recent edition of a popular technical writing textbook (Pauley & Riordan, 1990). A typical assignment from this book asks students to "Find an ad for a position in your field. . . . Based on the ad, decide which of your skills and experiences you should discuss. . . . Then write a letter . . . applying for the job" (p. 413).

Here, certainly, we find a recurring rhetorical situation. Students themselves are asked to search out examples of the situation, and they are asked to examine their own backgrounds, to locate particular experiences and skills from that background that might be applied as they respond to the job ad. There is value in using such typical rhetorical situations, especially in technical and professional writing courses where real-world writing is encouraged.

Problems begin to arise, however, when we look more closely at what the authors of this text ask students to do as they carry out this assignment. Here is a shortened version of a checklist they have provided for students to help them carry out the assignment:

- State the job you are applying for.
- State where you found out about the job.
- List what the employer needs.
- List your own skills, courses, or jobs.
- Write at least three paragraphs:
  —an introduction
  —a body
  —a conclusion (Pauley & Riordan, 1990, p. 409)

This checklist and the brief chapter on job application writing that sur-rounds it (pp. 402-413) provide basic examples of what often happens when genres are used in mechanistic ways, as templates into which stu-dents are encouraged to fit the facts and details of their experiences or reading. This listing of conventional rhetorical moves is supplemented by two sample job application letters that encourage students to look for text-ual features that they can then imitate in their own letters.

Rhetoric, in this case, simply means reducing the recurring situation (job-application writing) back to the most general and mechanical of exigencies: the issue of how the applicant's experience might be broken down into a conventional list of required skills and experience and how those experiences and skills can be fitted into the introduction, body, and conclusion format conventional to job applications.

But the perspective discussed here on genre as a process through which writers interpret context and develop strategies for approaching, analyzing, and responding to situations would require that the mecha-nistic rhetorical perspective applied in this text be subjected to further, more critical analysis and action. A pedagogy including this more bal-anced perspective would include questions such as the following: What are the connections between the general exigency described here as applying to all job application letters and the particular exigency apply-ing to the ad you, as a writer, have chosen? What kind of information could you bring to bear on this more particular exigency? How might you develop a strategy that will enable you as a writer to encompass both the general and particular aspects of this rhetorical situation? How might this strategy help you manage the choices you make as you compose your letter? How might these choices offset your treatment of substance, arrangement, style, and supporting material? Through what means (conversation, collaboration, negotiation, mediation, research, personal reflection) might you be able to find useful answers to the above questions?

These five critical questions merely begin to suggest the importance of full rhetorical analyses and strategy-construction to the process of composing. But they serve to close this essay with a brief and simple example of the need for a more complex and dynamic theory of genre in any writing course in which dialogic literacy is the primary goal.

Notes

1. Moffett (1968, 1981a, 1981b) also significantly contributed to the theoretical basis for the expressionist movement in the 1970s and 1980s in the United States. Both Moffett's and Britton's work were actually as much social constructionist as they were expressionistic. Both simply focused the work of Vygotsky and other social theorists on the writer engaged in the process of composing rather than on the rhetorical implications of that process. A recent symposium on Britton's work in *College Composition and Communication* attempted to correct current social constructionist misreadings of his contributions to composition theory and pedagogy (Tirrell, Pradl, Warnock, & Britton, 1990).

2. Fulkerson (1990), in an article updating earlier work on competing theories of composition, argues that recent combinations of rhetoric and social constructionist perspectives have led to the predominance of rhetorical emphases on readers and social contexts in composition theory, at the expense of expressive and formalist theories. Fulkerson seems, however, to be totally unaware of the fact that a great deal of recent literacy research supports the breaking down of the barriers between these now outdated ways of categorizing composition theory and practice.

3. Carter (1990) offers an insightful perspective on the extremely binary quality of the composition field's recent treatment of the opposition between the cognitive and social theorists. He calls for a balanced sociocognitive perspective as the most appropriate way to approach complex acts of composing. Carter's article indirectly addresses the opposition between expressive and social theories as well.

4. A recent issue of the *Association of Departments of English Bulletin* (Fall 1990) provides a clear indication that even very specialized discourse communities are questioning the hegemony of particular generic conventions. Articles by Marius (1990), Torgovnick (1990), and Atkins (1990) all in one way or another argue that strict academic exposition is and should be giving way to more plural and heteroglossic texts in English studies. These mixed or plural forms would better combine personal and professional meanings, these authors argue.

5. D. Sullivan (1990), in a recent *Journal of Advanced Composition* article, equates this need to mediate genre and choice in the act of composing with the classical concept of *kairos,* which he defines as the writer's ability to recognize and act on "the opportune moment" for the exercising of individual choice (p. 384).

6

You Are What You Cite:
Novelty and Intertextuality in
a Biologist's Experimental Article

CAROL BERKENKOTTER

THOMAS N. HUCKIN

> Intertextuality: Evidence, in a book, that the author has read some other
> book and remembers it.
>
> —Scanlan (n.d.)

At one bench in June Davis's laboratory, Beverly Cronin, a doctoral student, is busy cutting off the tips of mouse tails to draw blood for an assay.
It is a delicate procedure: To get the mice to hold still, she must lure them
into a test tube with food. Once there, they are secured and clipped at the
end of the tail. At another bench in the long rectangular room, Davis does
a cardiac puncture to draw blood directly from the hearts of the mice she
is studying. She selects one anesthetized mouse at a time from a jar

AUTHORS' NOTE: We wish to acknowledge, with appreciation, the constructive criticism
that Charles Bazerman, Alan Gross, and Donald Rubin provided us during the writing of this
chapter.

containing CO_2 and quickly makes an incision into its chest cavity. Pulling the ribs back, she exposes the heart of the animal to draw the blood with a small syringe. The samples taken from the tail clips and the heart are then placed into a centrifuge.

Two experiments are being conducted in Davis's lab this morning. Cronin is investigating the changes in blood serum for plasma fibrinogen, a clotting agent. She spreads a small amount of blood, to which sterile saline has been added, on the surface of an agar plate to determine whether the plate will grow colonies of *Candida albicans,* a common fungus that has been implicated in toxic shock syndrome (TSS). She is performing an assay that involves coating and washing 96 tiny plastic wells and then filling them with mouse serum, which is to be examined for the presence of a powerful monokine, tumor necrosis factor (TNF), which the body produces in response to disease producing agents.

Much of Davis's time is spent in the lab with her graduate students injecting mice with various infectious agents and toxins, drawing endless blood samples from an endless supply of mice, writing down different times and dosages in her lab notebook, performing assays and other procedures, doing statistical analyses of the data, looking for significant results, being disappointed, trying again. Davis is one of many scientists engaged in what might best be described as "brickwork," the routine tasks that constitute much of what Kuhn (1970) calls normal science. The tedium and monotony of this labor is considerable.

Such is life in the laboratory, a setting in which many scientists carry out the archetypical experiential activity (Bazerman, in press) of conducting experiments. Scientists engage in this activity to play their role in carrying out what most believe to be a global, rational enterprise. Lewontin (1991) in a recent article suggests:

> Most natural scientists are really positivists. They rely heavily both on confirmation and falsification, and they believe that the gathering of facts, followed by inference rather than the testing of theories, is the primary enterprise of science. *They are daily reinforced in their view of science by reading and writing the literature of science.* . . . Science consists, in this view of the postulation of more or less general assertions about causation and the necessary interconnection between repeatable phenomena. These postulations demand the gathering of facts: observation from nature or from the deliberate perturbations of nature that are called experiments. When the facts are in, they can be compared with the postulated relations to confirm or falsify the hypothetical world (p. 141; italics added)

Much of the literature of science is in the form of the four-part experimental article with its "Introduction," "Material and Methods," "Results," and "Discussion" sections, each of which serves to codify scientific activity into a coherent narrative of inductive discovery (see Medawar, 1964). This is to say that the four sections function together to present a rational view of scientific activity as a cumulative enterprise, building on accepted wisdom yet, at the same time, constantly seeking new knowledge. [1]

Our purpose in this chapter is to examine a particular narrative mechanism—the use of citation, or referencing—that is central to the generic function of the experimental article. Citation establishes the intertextual linkages that diachronically connect scientists' laboratory activity to significant activity in the field, and thus serves to establish a narrative context for the study to be reported. [2] The use of citations is intrinsic to scientists' story making because it contextualizes local (laboratory) knowledge within an ongoing history of disciplinary knowledge making. Such contextualization is essential because it is only when the scientist places his or her laboratory findings within a framework of accepted knowledge that a claim to have made a scientific discovery— and thereby to have contributed to the field's body of knowledge—can be made.

The concept of *novelty,* as it relates to scientific discovery, refers to the idea that innovations (new postulations) are at the heart of the scientific enterprise as it is seen by its practitioners. If scientific activity is to be purposeful and cumulative, then a major criterion for publication is the novelty or *surprise value* (Huckin, 1987) of the researchers' knowledge claims, seen in the context of accumulated knowledge. As Amsterdamska and Leydesdorff (1989) suggest:

> In a scientific article "the new encounters the old" for the first time. This encounter has a double significance since articles not only justify the new by showing that the result is warranted by experiment or observation or previous theory, but also place and integrate innovations into the context of "old" and accepted knowledge. . . . References which appear in the text are the most explicit manner in which the arguments presented in an article are portrayed as linked to other texts, and thus also to a particular body of knowledge. (p. 451)

From the above perspective, experimental articles that are deemed by journal reviewers as novel or "newsworthy" (see Huckin, 1987) have been positioned by their authors within an intertextual "web" (Bazerman, 1988) or "fabric" (Amsterdamska & Leydesdorff, 1989).[3] The linguistic/rhetorical strategies for integrating innovation (new knowledge claims) into the existing knowledge structure of one's field must be, as Amsterdamska and Leydesdorff (1989) propose, both implicit and explicit, diffuse and specific. This is to say that

> in the most diffuse manner, but perhaps most importantly, [integration] takes place through the shared technical and "theory-laden" language and through shared patterns of argumentation. A more specific "integration" of the new claim occurs when the use of a particular concept or method is said to be warranted by reference to some precedent, a previous occasion or context in which it has been used. In the Introduction and Conclusion sections of an article, this "placement" of a claim takes the form of a specific requirement that the problem which is to be addressed, its significance, and sometimes also its implications, be specified and explicitly stated. In a somewhat more implicit manner, such integration also takes place in the other sections of the article. Thanks to this integration, the innovation—no matter how trivial, or how original—is not just another loose fact added to the heap but rather an extension of a thread, a new knot, a strengthened connection, or alternatively a bit of unraveling, an indication of a "hole," a bit of reweaving, etc. (Amsterdamska & Leydesdorff, 1989, p. 451)

The study described in this chapter documents this process of integration of new claims into existing knowledge. Specifically, we focus on Davis and Cronin's efforts to get a manuscript accepted for publication by a major journal in Davis's field, *Infection and Immunity*. In our research we used a combination of sources for obtaining data, including field notes from lab observations, taped interviews (which took place between February 1988 and August 1990) with Davis (the senior researcher) and Cronin (a doctoral student writing her dissertation), the drafts of a single paper written primarily by Davis, the comments on those drafts by the author (made in interviews), the reviewers' written comments, the editors' written comments, and the authors' written responses to the editor. To understand in greater detail the kinds of responses a reviewer might make when reading the drafts of this paper, we also used three "penumbral readers," active researchers in fields close to but not identical to that of the author.[4] These researchers agreed to read both the first draft and final draft of Davis's paper as reviewers, commenting on audiotape as they read. We call these readers "penumbral" because we see them not as part of the writer's immediate target

audience (i.e., specialists in *Candida albicans*) but as typical of the
larger readership for her paper: Grant reviewers, manuscript reviewers,
and colleagues in related fields.

We examine the way that Davis and Cronin—at the most abstract and
diffuse level—use shared patterns of argument in the Introduction and
Conclusion sections of their article; we examine their reference to a prior
series of experiments reported in the journal *Science,* in 1985 and 1986,
the methodology and conceptual structure of which they have repli-
cated; finally, at the most explicit and specific level of argument, we
examine their use of citations to demonstrate how they position their
experiments in relation to other investigations of a related series of
problems on which scientists at the research front of their speciality are
working. This study of the peer review process uncovers the highly
contingent and tentative epistemological status of the scientist's knowledge claim. Observing what goes on in the course of peer review clearly
demonstrates the socially constituted, negotiational character of a genre
that has been most often analyzed only in the form of finished products
(see Gross, 1990). Although something can be gained by studying published reports, certainly, we feel that tracking and analyzing the development of a report as it goes through various revisions yields unique
insights about the epistemology of science.

Davis's Research Program

For more than a decade Davis has studied the role of *Candida albicans,*
a common yeast, in the mouth, intraperitoneal cavity, and vaginal tract.
Although perceived by the medical community as a benign nuisance,
C. albicans has been implicated in toxic shock syndrome, a life-threat-
ening disease occurring primarily in women. In the early 1980s Davis's
research suggested that *C. albicans* acted synergistically with *Staphylococcus aureus* bacteria, producing toxic effects in mice identified as
having TSS. Since that early research she had moved on to study the
effects of *C. albicans* on the immune system, having focused specifically on its relation to tumor necrosis factor. Immunologists have been
interested in studying the mechanism of TNF production because it is
one of the body's mediators when the organism is attacked by infectious
agents. As an immune cell, TNF produces potentially lethal effects. The
majority of *in vivo* TNF studies have been on responses to invasion by

endotoxin (a toxin that is released from certain bacteria as they disintegrate in the body, causing fever, shock, etc.) that can indirectly produce widespread tissue damage by generating an exaggerated and potentially lethal response from immune cells. A study published in *Science* in 1985 first established the fact of TNF's lethal overreaction to endotoxin invasion. The authors of this study noted that TNF functioned

> as a hormone to promote cellular responses which, in part, result in the mobilization of host energy reserves in response to invasion . . . [thus] in the present study we reasoned that TNF might also play a role in the lethal metabolic effects of endotoxin mediated shock. Accordingly, we passively immunized mice with antibody to TNF and challenged them with lethal amounts of [endotoxin] (Beutler, Milsark, & Cerami, 1985, p. 869).

Beutler et al. (1985) reported that when immune serum was administered to mice by "intraperitoneal injection 1.5 hours before the intraperitoneal injection of 400 μg of [endotoxin], a significant protective effect was demonstrable . . . compared to the mortality rate observed among control mice treated with preimmune [nonreactive] serum or with serum from nonimmunized rabbits" (p. 870). Although it established a cause-and-effect relationship between endotoxin and TNF, Beutler et al.'s study did not create such a relationship between external infectious agents and TNF. Hence, this was a natural direction for researchers to take, using infectious agents like *C. albicans*. In 1988, a second group of researchers reported that "*C. albicans* induced TNF production *in vitro* by human monocytes and natural killer cells" (Djeu, Blanchard, Richards, & Friedman quoted in Riipi & Carlson, 1988, p. 1). By the late 1980s Davis had become interested in determining the role *C. albicans* played in elevating the fibrinogen level in the blood, one of the effects of TNF. Davis and Cronin at that time had been replicating Beutler et al.'s *in vivo* study, examining the *C. albicans*/TNF relationship. They sent a report of their initial findings to the journal *Infection and Immunity,* noting that prior work on the *C. albicans*/TNF relationship had been done *in vitro,* and that theirs was the first *in vivo* study.

Constructing an Argument for Novelty

The particular rhetorical problem that Davis and Cronin faced was creating sufficient surprise value or novelty for journal reviewers to concur that their experiments were indeed newsworthy. Because the stuff of their science is brickwork rather than the *big ideas* that challenge the status quo (see Myers, 1985, 1990), their argumentative strategies were necessarily quite different. For Davis and Cronin, the rhetorical task was to justify the importance of their experiments by creating a research space in their discussion of recent work. That is, they needed to show in the Introduction and Discussion sections of their article what had been accomplished in the recent research on TNF and endotoxin and to propose in what ways their experiments extended this line of research on the production of TNF in mice by infectious agents. The importance of positioning their work in relation to a series of crucial experiments, i.e., deploying the relevant line of research through citation, cannot be underestimated. As one of our penumbral readers observed:

> In a good Introduction you can almost get the entire paper and background. A good introduction, in a way, is like Kentucky Fried Chicken: the Colonel used to say, if the gravy is good enough you can throw away the chicken. And you can almost throw away the paper if the Introduction is set up well, because you can see the field, you can see where it all fits, you can see what they did. (JW)

Draft 1: Getting the Basic Facts Down

Davis (who as senior researcher wrote the entire paper except for the Methods section) did not, however, seem concerned about laying out the field and positioning her work within it, at least not initially. In draft 1, the Introduction is local rather than intertextual, with the author citing only her own studies:

> We have previously found that a small dose of *Candida albicans* which had little adverse effect by itself, acted synergistically with *Staphylococcus aureus* to cause shock and death in mice (3 [self-citation]). While attempting to identify the role of *C. albicans* in the *C. albicans/S. aureus* synergism, we had found that *C. albicans* alone at low doses which have no effect on a variety of blood parameters tested did elevate plasma fibrinogen levels (unpublished). In this study the ability of small doses of *C. albicans* to induce changes on blood chemistry and hemotology and the role of tumor necrosis factor (TNF) in these changes were examined. (Davis, draft 1)

The Introduction is curiously insular in this first draft. Davis does not situate her work within any body of experiments conducted by other specialists. Rather, she attempts only to link the current study to her own previous research on synergism between *Candida albicans* and *Staphylococcus aureus,* specifically to follow up on one result in a previous (unpublished) study that *C. albicans* induction had produced elevated fibrinogen levels in the blood.[5] Although she mentions TNF in this early Introduction, its significance in being induced by *C. albicans* is only alluded to in the last sentence. Reference to the experiments by Beutler, Milsark, and Cerami on the production of TNF by endotoxin is made only once, buried in the second paragraph of her Discussion section: "Endotoxin, cause of endotoxic shock, has been shown to induce TNF which is in turn responsible for shock in the mouse (2)."

Draft 2: Building the TNF Connection

Although Davis was not unaware of the importance of positioning her work within a related literature, during the revising of the first and second drafts she was much more concerned with technical issues and questions that reviewers raised about the procedures used in the lab. For example, one reviewer wanted to know if precautions were taken to guard against endotoxin (LPS) contamination of *C. albicans* preparations and other reagents used. She or he felt this information to be important since even picogram amounts of LPS could act in synergy with *C. albicans,* thus contributing to the observed responses. Couched in the polite and cautious language of the scientist persona, the register masks the somewhat hostile implication of the question, that procedural sloppiness may have affected what was observed. The comment can thus be seen as an indirect request that Davis produce evidence showing that such was not the case. Responding to this and other questions of lab procedure occupied Davis's efforts through her first few revisions.

At the same time, the Introduction to draft 2 possesses new intertextual features. For one thing, Davis substitutes a reference to the work of Beutler et al. in place of the reference to her own unpublished study on *C. albicans* and fibrinogen (see sentence 3):

> We have previously found that a small dose of *Candida albicans,* which had little adverse effect by itself, acted synergistically with *Staphylococcus aureus* to cause shock and death in mice (3 [self-citation]). This study was undertaken to determine how *C. albicans* contributes to this lethal shock

> synergism. It has been reported that induced TNF is responsible for endo-
> toxic shock in the mouse (2 [reference to Beutler et al.]). Because *C.
> albicans* and endotoxin share a number of characteristics (for review see
> 12) the role of TNF in candidal-induced hematology and blood chemistry
> changes was examined. (Davis, draft 2)

Not only does Davis link her study to the experiments of Beutler et
al. but she also inserts another intertextual reference in sentence 4. In
this sentence she also begins to develop what one of our penumbral
readers described as a "prospective rationale" in the form of a general
warranting statement implied by her assertion. The warrant implied
in sentence 4 can be seen to have the form of the following heuristic
reasoning: If two chemical substances x and y share characteristics A
through M, and if x is then found to possess the characteristic N, it is
reasonable to suppose that y might also possess N.

This warrant is not made explicit, probably because everyone in Davis's
discourse community would already subscribe to it. Davis's claim that
C. albicans shares characteristics with endotoxin is, in this sense, in-
dexical, pointing as it does to a shared warrant within the field. This
kind of logical appeal, we suggest, is based on tacit presuppositional
knowledge within a discourse community. What is being warranted is
Davis's assertion that a yeast and the cell wall of a bacterium have similar characteristics, an assertion that our penumbral readers questioned:

> When . . . they say something like "*albicans* and endotoxin share a number
> of characteristics," what are those characteristics? I don't know what they
> are. And it doesn't do me any good to tell me to go see a review because
> the last thing I'm going to do is go look up a review to find out what the
> characteristics are, to understand the intellectual linkage that these things
> have. (JW)

> I have no idea what they mean by *Candida albicans* and endotoxin sharing
> a number of characteristics. The response of an animal to *Candida albicans*
> and endotoxin may share a number of characteristics, but *Candida albicans*
> and endotoxin don't share anything. Very different stuff. (BA)

These two penumbral readers may have accepted the implied warrant but
they rejected the specific grounds of the claim (i.e., that *C. albicans* and
endotoxin share a number of characteristics). Davis's inclusion of a citation
to an article that purportedly demonstrates a connection between the two

indicates her awareness of the need to provide such grounds for skeptical or uninformed readers.

Draft 2 was rejected by the editor of the *Journal of Medical and Veterinary Mycology,* although, as Davis pointed out in an interview, the reasons for rejection were most likely due to the lack of "interesting" data on the *C. albicans*/TNF relationship. What she meant by this observation is that the major newsworthy item was her inference that "TNF was induced in the mouse by *C. albicans* infection, and that this TNF in turn, was responsible for the observed increase in fibrinogen levels." This claim, however, is based on *indirect evidence* (the presence of elevated levels of fibrinogen in the mice not treated with TNF antibodies 18 hours before injection with TNF) rather than on direct observation. The reviewers found this inference alone to be insufficient to support Davis's claim regarding the *C. albicans*/TNF relationship. Thus the editor rejected the paper on the grounds that it was incomplete, lacking data on TNF.

Draft 3: Providing the Necessary Data on TNF, Linking It to *Candida Albicans*

Eight months and dozens of experiments later, using a new assay that enabled them to report the levels of TNF produced by injections in mice with *C. albicans,* Davis and Cronin were able to produce quantitative data that directly supported their claims that *C. albicans* induces TNF in mice and that TNF was responsible for an increase in fibrinogen. Davis (1990) reported this information in a letter to the editor of *Infection and Immunity* to whom she sent her revision:

> We have included new data on candidal-induced TNF using a sandwich type ELISA [enzyme-linked immunosorbent assay]. . . . As you stated that you would be glad to look at this paper again should we provide this additional data, we are returning it to you with the hope of a better outcome. Specifically we have . . . determined with ELISA the levels of circulating TNF after selected doses of *C. albicans.* (letter from June Davis to editor of *Infection and Immunity*)

To reflect their emphasis on the new data and especially to foreground the newsworthy finding that TNF could be induced by *C. albicans* (a yeast) as well as by endotoxin (cell walls of bacteria), the authors changed the title of their paper from "Elevation of Fibrinogen Levels

Due to Injection of *Candida albicans* or Recombinant Tumor Necrosis Factor (TNF) in the Mouse: Protection with TNF Antibodies" to "Tumor Necrosis Factor (TNF) Is Induced in Mice by *Candida albicans:* Role of TNF in Fibrinogen Increase." Not only does the second title foreground the *Candida*/TNF relationship but it does so very emphatically by putting it in the form of a full predication, as is done in newspaper headlines (see Huckin, 1987, for further discussion of this linguistic feature). Furthermore, Davis and Cronin changed the wording of sentence 4 in the Introduction somewhat to highlight their new central claim that *C. albicans* induces TNF production. Instead of saying "the role of TNF in candidal-induced hematology and blood chemistry changes was examined," they say, more directly, "candidal-infected mice were examined for induced TNF."

Both reviewers for *Infection and Immunity* approved of the direction the paper was taking. But they also had suggestions for further textual/rhetorical improvements, such as expanding the Introduction. One reviewer had indicated that the new draft was on much firmer ground than the manuscript he or she had previously reviewed. But this reviewer had also submitted a checklist of several items that needed to be corrected. These included expanding the Introduction to set the stage for the research that followed (the reviewer was bothered that the readers unfamiliar with TNF and fibrinogen relationships would not understand why fibrinogen was selected for assay). The reviewer also wanted to see included in the introduction the information buried in the second paragraph of the discussion. The other reviewer concurred, suggesting that further references to related studies (which had not appeared in the manuscript) also should be included in the Introduction, in particular the results of previous studies on the induction of TNF by *C. albicans* and the ability of TNF to potentiate the anticandidal activity of polymorphonuclear leukocytes, as reported in the *Journal of Immunology*, 1986, and *Infection and Immunity*, 1989. This reviewer suggested a further intertextual move by exhorting the investigators to be explicit with their finding that the monokine response of mice to *C. albicans* differs from that obtained with LPS [endotoxin]. The reviewer added that these differences are important and should be addressed.

Draft 4: Creating the "Phony Story"

Davis immediately saw the validity of the above reviewer's comment about monokine response[6] and set about doing "a thorough comparative

study of the TNF responses to endotoxin and *C. albicans*," as she was
to report to the editor of *Infection and Immunity* in the letter that ac-
companied her revision. On the rhetorical level, though, she was much
less enthusiastic. Preparing to write the next draft of the manuscript,
Davis alluded to the "phony story" that she would have to add to the
Introduction. When pressed as to what she meant, she simply noted that
"reviewers always expect you to say certain things." Davis's efforts to
develop a phony story, i.e., to contextualize her experiments within a
related literature can be seen in draft 4. First, in the Discussion section
she adds two paragraphs in which she compares her *C. albicans*/TNF
data with that of Beutler et al. as well as the data from a second group
of researchers (Zuckerman & Bendule, 1989) who had measured TNF
levels in mice receiving lethal doses of endotoxin. Second, she expands
the citational base of the Introduction, incorporating references the reviewers had suggested and using accepted knowledge claims as scaffolding for the present study. An examination of the Introductions to
drafts 4 and 5 will illustrate the ways in which Davis integrates the new
with the old.

> We have previously found that a small dose of *Candida albicans* which had
> little adverse effect by itself, acted synergistically with *Staphylococcus aureus*
> to cause shock and death in mice (3). The present study was undertaken to
> determine how *C. albicans* contributes to this lethal shock synergism. It
> has been reported that induced tumor necrosis factor (TNF) is responsible
> for endotoxic shock in the mouse (2). Because *C. albicans* and endotoxin
> share a number of characteristics (for a review see 13) candidal-infected
> mice were examined for induced TNF. As exogenously administered TNF
> is known to induce acute phase proteins such as fibrinogen (9), plasma
> fibrinogen in infected mice was also measured and the role of TNF in the
> fibrinogen increase investigated.
>
> It has been reported recently that *C. albicans* induced TNF production
> by human monocytes and NK cells *in vitro* (8). Also *in vitro* TNF has been
> shown to potentiate the fungicidal activity of human neutrophils against
> *C. albicans* (9, 10). (Davis, draft 4)

Davis made three major additions in this draft of the Introduction.
The first of these, the last sentence in paragraph 1, was constructed in
response to the comments of the first reviewer, who had wanted to see
information originally appearing in the Discussion foregrounded in the
Introduction and who was also concerned that "the reader who is not
familiar with TNF and fibrinogen relationships does not have a clue as

to why fibrinogen was selected for the assay." To accommodate this reviewer, Davis combined two sentences from the previous draft. She revised the sentence from the Introduction in draft 3—"In addition the relationship between TNF and candidal-induced hematology and blood chemistry changes was examined"—and combined it with a sentence containing a citation that had appeared originally in paragraph 2 of the Discussion section of draft 3: "Also, exogenously administered TNF is known to induce acute phase proteins such as fibrinogen (9)." In the revised sentence, one can see Davis building her study on an antecedent, and thus linking new to old, or given, knowledge. This kind of linking continues to supply a prospective rationale, justifying the present study with what has previously been accomplished in the field. This rationale creates the appearance of a chronology of scientific activity; one can therefore understand why Davis thought of herself as constructing a phony story. She was, in fact, constructing a narrative. Davis also added two new sentences that contained citations in a very brief second paragraph. "It has been reported recently that *C. albicans* induced TNF production by human monocytes and NK cells *in vitro* (8). Also *in vitro* TNF has been shown to potentiate the fungicidal activity of human neutrophils against *C. albicans* (9, 10)." These sentences are included to accommodate the second reviewer, who had suggested that she add references to two *in vitro* studies in the Introduction and who had given her the citation for one of these, the other having been buried in the Discussion of Davis's previous draft.

Draft 5: Strengthening the Argument

In the Introduction to draft 5, these two sentences (sentences 5 and 6) are skillfully woven into the body of the first paragraph:

> We have previously found that a small dose of *Candida albicans* which had little adverse effect by itself, acted synergistically with *Staphylococcus aureus* to cause shock and death in mice (3). The present study was undertaken to determine how *C. albicans* contributes to this lethal shock synergism. It has been reported that induced tumor necrosis factor (TNF) is responsible for endotoxic shock in the mouse (2). Because *C. albicans* and endotoxin share a number of characteristics (for a review see 15) candidal-infected mice were examined for induced TNF. It is reasonable to suspect that *C. albicans* could induce TNF *in vivo* because it has been reported recently that *C. albicans* induced TNF production *in vitro* by human

monocytes and natural killer cells (9). TNF has also been shown to potentiate the fungicidal activity of human neutrophils *in vitro* against *C. albicans* (10, 11). As exogenously administered TNF is known to induce acute phase proteins such as fibrinogen (12), plasma fibrinogen in infected mice was also measured and the role of TNF in the fibrinogen increase investigated. (Davis, draft 5)

Embedded at the presuppositional level in these two sentences is a warrant in the form of a tacit methodological principle: If *a* produces *b in vitro*, then it is reasonable to suppose that *a* will produce *b in vivo* (other things being equal). This warrant underlies both the claim ("it is reasonable to suspect that *C. albicans* could induce TNF *in vivo*"—justification of the present study) and the grounds for that claim ("because it has been reported recently that *C. albicans* induced TNF production *in vitro* by human monocytes and natural killer cells [9]" and "TNF has also been shown to potentiate the fungicidal activity of human neutrophils *in vitro* against *C. albicans* [10, 11].") The warrant itself is not explicit; however, once again, as in sentence 3, biologist readers are likely to infer it. The references further buttress the claim by providing evidence and by being attached (by inference) to the underlying methodological principle. With its claim/grounds/warrant structure, sentence 5, like sentence 3, appears to fulfill the criteria for a Toulmin (1983) argument. Sentences 4 through 6 appear to exemplify two kinds of intertextual integration that Amsterdamska and Leydesdorff (1989) describe (although they do not use this term) when they refer to diffuse "shared patterns of argument" and the more specific integration of a new claim that occurs "when the use of a particular concept or method is said to be warranted by reference to some precedent, a previous occasion or context in which it has been used" (p. 451).

Looking at the Introduction as a microtext, we can trace a chronology, or story, that weaves together Davis's previous work with *C. albicans* and *S. aureus* synergism that produced shock in mice, with related sets of experiments that investigated the role of the immune system in response to infectious agents: *in vivo* studies that investigated the role of TNF production in endotoxic shock, *in vitro* studies of *C. albicans*-induced TNF production, and experiments that induced the *in vitro* production of acute phase proteins by exogenously administering TNF. Bound together with citations and undergirded by presuppositional warrants, these sentences create a coherent narrative that functions to contextualize (and thus justify) the present study. The creation of this kind of narrative scaffolding is quite different from the "create a research space" rhetorical activity that Swales (1990) describes.

Penumbral Readers' Perceptions of Novelty in Davis's First and Last Drafts

In this discussion of the revisions of Davis and Cronin's paper for the journal *Infection and Immunity,* we have focused on Davis's interactions with her editor and reviewers concerning the placing and integrating of her research into a related body of literature. One of our penumbral readers' responses to the first and final (published) versions of the manuscript illustrate how the scientist-reader looks for and responds to "news." Reading the first draft of the manuscript, JW said:

> To me this is a rather mundane finding, but I have to spend more time reading it to see whether or not it's really unique in terms of science, or whether or not it's merely continuing what you'd expect. If you inject TNF into animals, I guess I'm not surprised that a lot of things happen, cause it's nasty stuff. That's usually where the Introduction helps me a lot. They don't help me here.

Later, while reading the Discussion section of the final version in manuscript form, he commented on what caught his eye and interest:

JW: So now we get through everything and go to the Discussion. [Reads.] They're giving me their data and giving me their interpretation. [Reads.] They say this? I certainly didn't catch it [in the earlier draft]. [Reads.] Hmm. I don't see it. Starting this paragraph, [reading aloud] "Since endotoxin does not induce TNF in the C3 H/HeJ [endotoxin-resistant] strain, our finding that *C. albicans* did induce TNF in the endotoxin resistant strain suggests that the induction of TNF by *C. albicans* is under a different mechanism of control than the induction of TNF by endotoxin." Now if they said that in here (draft 1), I didn't even see it. But that's very interesting.

Interviewer: Why is that particularly interesting?

JW: It gives me another pathway. They're now defining something which is novel to me. They're not saying that the

induction of TNF by *Candida* is the same thing as the
endotoxin-induction of TNF; they're saying that it's a different mechanism, because of the [prior] work on endotoxin-
resistant strains. So, uh, that's telling me that this is new
and novel. So if I'm reviewing it, they get a little click on
the new and novel scale.

JW's reaction as he read the first and last drafts of the manuscript (as
well as our other readers' responses) provided important information
regarding the ways that other biologists, and specifically immunologists,
might have evaluated both the technical/scientific and the rhetorical
dimensions of this experimental report. These responses make us somewhat cautious about framing our discussion of novelty strictly within a
rhetorical context, something that researchers in our field are likely
to do. On the other hand, it was also JW who provided the insight that,
when written well, the Introduction enables the scientist reader with
particular schemata and purposes for reading (Bazerman, 1985) to "see
the field, . . . see where it all fits, . . . see what they did."

Conclusion

The study reported in this chapter documents what happened in one
case in one small arena of textual activity intimately connected to other
domains of scientific activity: A scientist placing her work within an
intertext. In many respects, Davis strikes us as a "typical" working
scientist, one who sees laboratory research and rhetorical activity as
distinctly *separate*. Her distaste for the disingenuous, her cynicism
regarding the phony story she would have to construct, suggests that
she was very aware of the difference between recounting local history
in her lab and contextualizing that history within a narrative framework,
and that she considered only the former as constituting true science. She
appears to be a very technically-minded scientist who devalues the rhetorical dimension, letting her reviewers tell her what's needed to create the
necessary phony story. Although she was a realist, her realism extended
only as far as what she observed in her lab (unlike the more rhetorically
savvy scientists described in Myers, 1990).
    Davis's reviewers, on the other hand, made no such distinction
between laboratory activity and rhetorical accommodation, insisting

that to be "science" her report had to include an intertextual framework for her local knowledge. Rather than letting the laboratory research speak for itself, they helped inject surprise value into this article by insisting on background information and warrants and by asking the author to foreground her claims. In this sense it can be said that the reviewers aided Davis in the construction of knowledge, i.e., in the claim to novelty.

Another framework for characterizing the reviewers' intellectual contribution is that of current narrative theory (Fisher, 1987), particularly in regard to *narrative coherence,* or the sense of completeness that a good story has. Davis's initial accounts of her research depict it as the story of her work in the lab: She obtained endotoxin-resistant mice, prepared *albicans* cultures, inoculated the mice, took blood samples, and so on. It is a distinctly "local" narrative. But Davis's reviewers have a much larger narrative in mind, the ongoing narrative of tumor necrosis factor and its role in fibrinogen production. They know that this larger narrative already commands the attention of a certain segment of the scientific community, and they can see how Davis's local narrative fits within it. Far from being a phony story, as Davis puts it, the larger narrative is, in a sense, the *real* story. At least, that is how the reviewers (and our penumbral readers) saw it.

A major question that we are left with is whether Davis's negative attitude toward writing phony stories and providing "the things that reviewers expect you to say," doesn't capture something more subtle about the ideological nature of the experimental article as a genre. This something more subtle has to do with the difference between Davis's perception of science and that of the reviewers. As we have said, Davis's view of science appears to be constituted largely by what goes on in the lab. In contrast, the perspective promoted by the reviewers, which sees laboratory activity in terms of its relevance to disciplinary intertextuality, functions to instantiate a rationalist view of the cumulative nature of scientific discovery that goes well beyond Davis's straightforward empiricism. This conception of science as an inductive, cumulative activity is what Lewontin (1991) seems to be referring to (in the passage we quoted earlier) when he talks about scientists being "daily reinforced in their view of science by reading and writing the literature of science" (p. 141).

From an entirely different perspective, Amsterdemska and Leydesdorff (1989) have hinted (although nonreflexively) at the ideological activity underlying scientists' conception of novelty:

Thanks to this integration [into a related literature], the innovation—no matter how trivial, or how original—is not just another loose fact added to the heap, but rather an extension of a thread, a new knot, a strengthened connection, or alternatively a bit of unraveling, an indication of a "hole," a bit of reweaving, etc. (p. 451)

The above view seems to reflect an assumption that is part of the methodological hard core (Lakotos, 1970) of scientists' beliefs concerning science as a rationalist enterprise.

However, we suggest that it is the unreflective use of the conventions themselves that reproduces this view of science. That is to say, the conventions of the scientific journal article instantiate ideological assumptions, assumptions that are regularly reinforced by scientists' routine, unreflexive use of the genre. As Bruner (1991) has argued, "Genres . . . are ways of telling that predispose us to use our minds and sensibilities in particular ways. In a word, while they may be representations of social ontology, they are also invitations to a particular style of epistemology" (p. 15). In the case of the experimental journal article, scientists who wish to have their work published must adopt a slightly contradictory stance in which they (1) act as though scientific discovery were a purely inductive process (see Medawar, 1964, pp. 9, 41) but (2) explicitly acknowledge, via appropriate citations and warrants, that hypotheses are inspired by earlier research done by other scientists.

We think that this study opens an interesting line of inquiry into the rhetoric of science regarding the relationship between the conventions of the genre and the ideological assumptions embedded in those conventions. What do the narratives that scientists are socialized to produce and which reify scientific activity as perceived by the community reveal about the social dynamics of such texts? What does a study of the ideological character of the genre's conventions reveal about the deeply embedded epistemological assumptions that permeate scientists' discursive practices in general? More broadly speaking, what can researchers in rhetoric and composition add to the provocative work in the rhetoric of science that draws on the rich interdisciplinary character of science studies to answer such questions? How do we draw on our own considerable expertise in analyzing the textual dynamics of discourse communities to begin to develop situationally located, "grounded" theories of genre? For rhetoric and composition specialists trained in English studies, the study of nonliterary genres has much promise both for research and for the teaching of academic writing across the curriculum.

Notes

1. This textual representation serves to instantiate a view of science that has been questioned and analyzed by a number of sociologists of science interested in contrasting formal with informal scientific discourse (see especially Gilbert & Mulkay, 1980, p. 198; Latour & Woolgar, 1986). These researchers argue that the rationalist view of scientific activity (as reproduced in the genre of the experimental article) brackets out the social and political variables affecting the laboratory decisions that scientists make.

2. The textual mechanisms through which scientists establish the intertextual linkages between their own and others' experiments has been investigated rather extensively by sociologists of science (Gilbert, 1977; Gilbert & Mulkay, 1980; Latour, 1987; Latour & Woolgar, 1986), by rhetoricians (Bazerman, 1988; Myers, 1990), by scientometricians who study citation patterns within various specialties (e.g., Amsterdamska & Leydesdorff, 1989; Cozzens, 1985, 1989; Leydesdorff & Amsterdamska, 1990; Small, 1977, 1978), and by citation content analysts (e.g., Swales, 1986) who study researchers' citing behaviors.

3. However, it should be noted that an intertext representing the work of a research network (or that of overlapping networks) is organized by problems at the research front—which are in turn affected by a specialty's reputational structure: who is being cited; who is being funded; and, therefore, who is seen as working at the field's cutting edge. Moreover, the intertext is subject to continual revision as each newsworthy article appears in print and is read and, in turn, cited by interested colleagues. Thus the textual matrix within which the activities of a particular community of scientists are communicated and acted on cannot be regarded as a static entity, but rather a set of dynamic relationships that are more or less stable depending on the community's cohesiveness.

4. One (BA) was a cellular immunologist specializing in the murine immune system, including TNF; another (KB) was a cellular immunologist specializing in the activation and regulation of T-cell responses; the third (JW) was a molecular immunologist/biologist specializing in mouse genetics.

5. Fibrinogen is a globulin, or protein, that is transformed into fibrin in the clotting of the blood. It is an important factor in heart disease. Because much of Davis's research has been supported by the National Heart Association, her interest in fibrinogen is understandable.

6. See Gross (1990) for a discussion of scientists' rhetorical strategies in responding to peer reviewers.

7

The Role of Law, Policy, and Ethics
in Corporate Composing:
Toward a Practical Ethics for
Professional Writing

JAMES E. PORTER

Research on what might be termed *corporate composing* looks at how writers in corporations interact to produce large-scale documents such as recommendation and feasibility reports, computer manuals, product information and packaging, brochures and quarterly reports, and advertisements (Doheny-Farina, 1986; Paradis, Dobrin, & Miller, 1985; Simpson, 1991; Spilka, 1988, 1990). Such research focuses on how documents within corporations are socially constructed (Bruffee, 1986; Faigley, 1985), that is, produced not simply by an independent and autonomous author writing to transfer information to a well-defined audience but through social interaction and negotiation of conflicting perspectives by numerous writers and editors producing documents to be read by diverse audiences. Another body of research focuses on the importance of ethical and legal factors in professional writing (Harcourt, 1990b; Johannesen, 1990; McCord, 1991; Pettit, Vaught, & Pulley, 1990; Porter, 1987, 1989, 1990b; Rentz & Debs, 1987; Scheibal, 1986; Speck, 1990; Stevenson, 1986; Walzer, 1989).

In this chapter I bring these two research interests together to explore the role of law, policy, and ethics in corporate composing and argue (1) that studies of corporate composing should consider law, policy, and ethics as important variables of workplace writing practice and (2) that treatments of ethics in professional writing should consider the social nature of ethical decision making. Typically, workplace studies of corporate composing practices do not say much about the ethical, legal, and policy constraints writers work under—and perhaps there is good reason why not: Often workplace writers are not conscious of these constraints. Meanwhile, studies of ethics in professional writing typically focus on ethical issues impacting the individual writer—and there is probably a good reason for this as well: The dominant assumption in the domain of ethics is that individuals make ethical judgments. The liberal humanist position that serves as academic caretaker of such matters assigns ethical responsibility to the solitary writer.

This chapter connects two bodies of research that have much of value to say to one another. My strategy in this chapter will be (1) to critique conventional treatments of ethics in professional writing, challenging their dominant critical assumptions; (2) to propose an alternative ethical approach, based on the notion of *praxis* and deriving from a social (and sophistic) approach to ethics; and (3) to demonstrate how law and policy can be invoked as concrete sources for this social ethic, an ethic that can guide corporate composing practice. Through this process I hope to establish the practical importance of ethics as a component of workplace writing.

Conventional Treatments of Ethics in Professional Writing

Discussions of the role of ethics in professional writing frequently appear in journals (like the *Journal of Business Communication*), and there is a well-established custom of treating ethical and legal matters in special journal issues or in bibliographies (Doheny-Farina, 1987, 1989a; *Journal of Business Communication,* 1990; Reinsch, 1990; Shirk, 1990; Speck & Porter, 1990). Those who write on the subject agree that ethics is an important though too-much-neglected topic, but paradoxically, these discussions have been based on a view of ethics that undercuts their aims and makes it all too easy for corporations, and perhaps textbooks as well, to ignore their admonishments. Generally, these discussions have treated ethics as a matter of individual decision making concerning infrequent (and, often, large-scale) problems, as separate from and prior to composing, and as determining composing practice.

Ethical Problems as Large Scale, Infrequent, and Individual

Although articles citing evidence of unethical corporate behavior frequently name corporations as perpetrators, the dominant assumption in discussions of business communication ethics seems to be that "companies don't commit unethical acts, people do" (Spencer & Lehman, 1990, p. 7). Consistent with this warrant (and perhaps also because education tends to focus on the individual learner), cases typically present ethical problems as decisions facing individual writers (see, for example, Hobel, 1989). In contrast, research on corporate composing indicates that for large-scale corporate documents (which are frequently produced through extensive collaboration), there might well be no one single person responsible for the ethical judgment. In fact, for some documents (e.g., advertising and product packaging), *the author is the entire corporation,* not any single writer within the corporation.

Some treatments portray ethical dilemmas as dramatic, large-scale, and infrequent problems. We are fond of citing cases that come to the public eye, often involving major life- or job-threatening situations requiring that the individual stand against or cooperate with the corporation. The infamous whistle-blowing cases such as the incident at the Dresden Nuclear Power Station, the *Challenger* space shuttle explosion, or the Three Mile Island incident (Fitzgerald, 1990) may come to mind. Discussions of these cases sometimes portray the ethical decision as an either/or proposition: Either you cooperate with the company (and hence risk behaving unethically), or you act like a hero and blow the whistle (and risk losing your job). Such cases are often post facto examples showing instances in which the individual was right to stand against the corporation—thus promoting a sense of ethics as individual heroism.

I submit that such narratives are not particularly appropriate for teaching the ethics of professional communication. Though these tales certainly have dramatic appeal, and for that reason command attention, such incidents also send the message that whistle-blowing incidents are infrequent (so ethics is something you may never have to worry about) and that being ethical means opposing the company when you know you are right. These conventional discussions of ethics arise from a notion of professional communication as information transfer (Driskill, 1989; Paradis et al., 1985): the writer serving as conduit (or bridge) channeling technical information to the ignorant audience (G. Clark, 1987). This communication model assigns inventional authority and ethical responsibility to the individual writer—though both social constructionist theory and workplace writing studies suggest that corporate composing is in the main collaborative.

Ethical Decisions Before Composing

Ethical discussions and cases also reveal a tendency, especially in business communication, to conflate *professional writing ethics* with *business ethics,* assuming that a more ethical approach to business generally will reflect itself in more ethical writing (see Hunter, 1990; Lewis & Speck, 1990; Spencer & Lehman, 1990). The assumptions here are that the ethical judgment is made before composing and that the act of writing is itself neutral, simply the means of implementing one's ethical (or unethical) decision. There is little sense that the composing act itself raises ethical concerns.

This view separates composing from ethics, placing them in separate compartments. Ethics and invention occur, and thus "truth" is established before composing. This compartmentalization explains the tendency to treat ethics as a component of style (by which is usually meant word choice) and to define an ethical style as one that accurately reflects the reality or truth of the situation. The most ethical rhetoric is assumed to be the plain, direct, unadorned style, which places the fewest language barriers between the reader and truth or reality.

It is no wonder that ethics has not been treated very seriously in professional writing textbooks, if it is not perceived as a topic having much to do with professional composition (though token treatment may be viewed as politically efficacious). This may also explain the common curricular practice of handling ethical matters in a short passage of a textbook or in a class period or two (see Golen, Powers, & Titkemeyer, 1985, p. 78; Harcourt, 1990a).

Ethical Theory as Determining Practice

Discussions of ethics in professional writing often proceed using a well-established analytic pattern: They construct or invoke some ethical principle (e.g., utilitarianism, Grice's maxims, Christian morality) and apply such theory to some problematic workplace situation to develop an ethical response to the problem, which the writer can then represent in a document (Yoos, 1984). This pattern participates in the popular binary privileging theory over practice, for which theory refers to general and/ or abstract designs, systems, beliefs, or principles used to explain, interpret, prescribe, or critique practice and practice refers to specific composing activities—i.e., to local and particular workplace writing situations (see P. Sullivan & Porter, in press).

Those who teach or do professional writing are very much aware of the ubiquity of the theory/practice binary. The theory/practice division may be a fictional construct—but it nonetheless has quite powerful effects. It leads *theoretical people* to privilege lofty, abstract, and (usually) static principles without sufficient regard for their dynamic *situatedness*. It leads *practical people* to neglect or underestimate discussions of ideology or theory as too abstract, too academic, too political—and as neither necessary nor desirable in the workplace.

Some in professional writing think that professional writing pedagogy has been too reliant on workplace research (Parsons, 1987). Others disagree, claiming that "the point of view of the business and industrial world of which the student will become a part is the only criterion which should be used to plan and teach" technical writing (Tebeaux, 1980, p. 823). However, the dominant assumption of those writing about ethics has been that practice must look to theory for ethical authority. This authority often takes the form of an appeal to an ethical tradition (e.g., Graeco-Roman philosophy or Judeo-Christian belief) found in historical texts (for example, see Lewis & Speck, 1990).

If we accept that professional communication is a process of social interaction, as the workplace researchers have been showing us, then we have to question some of the common assumptions about ethics and work to develop an ethic that accounts for the social nature of corporate composing. (Later in this chapter I attempt to construct such an ethic.) Such an ethic would not deny the validity of individual ethical responsibility—but it would insist that ethical responsibility, like a corporate document, is socially constructed and therefore must be shared. The other implication of this stance is that it views ethics not as an infrequent moral problem that may arise—not just as a matter of the occasional whistle-blowing decision—but as much more fundamental to communication. It is a perspective—or a pluralistic critique—that should be consciously invoked for every communication, because every communication influences social relations. Every instance of corporate composing has ethical as well as legal consequences.

Toward a Social and Sophistic Ethic

A number of researchers in a variety of fields have challenged the traditional academic privileging of theory over practice and argue for a balanced perspective that integrates theory and situatedness (Bourdieu, 1977; De Certeau, 1984; Geertz, 1983; Phelps, 1988; Suchman, 1987; P. Sullivan & Porter, in press; Winograd & Flores, 1986). In rhetoric and composition, C. R. Miller (1989) and Phelps (1988) have both recognized the limitations of theory and have worked to develop a dynamic *praxis* that recognizes the necessary contribution of practice and setting, or *kairos* (see Kinneavy, 1986). For example, Miller (1989) identifies *praxis* as a middle ground between theory and practice, a higher form of practice. *Praxis* is more than a simple addition of or compromise between theory and practice; it represents a new kind of critical positioning. It is a practice, conscious of itself, that calls on prudential reasoning for the sake not only of production but for right conduct as well. It is an informed and politically conscious action.

The judgment that enables *praxis* is practical wisdom—that is, *phronesis,* sometimes translated as "prudence" (see Aristotle, 1976, book 6). According to Garver (1987), prudence inserts itself into that gap "between apprehending a rule" [*episteme*] and applying it [*techne*]" (p. 16). Prudence "requires that the writer find some middle ground between too much universality—the superfluous . . . proclamation of moralizing principles—and too much particularity" (p. 39). Ethically, prudence is "halfway between an ethics of principles, in which those principles univocally dictate action . . . and an ethics of consequences, in which the successful result is all" (p. 12). Thus *praxis* is in this sense sophistic and contingent in nature.

A shift to *praxis* would have profound implications for teaching and research in professional writing as well as workplace writing practice. First, this changed positionality influences how we define and locate ethical authority. *Praxis* suggests that the source of ethics is not "pure practice" (if there could ever be such a thing): We should not simply observe corporate composing practices uncritically to determine strategies and principles appropriate for professional writing. Nor should we observe those practices simply to test the adequacy of a given theory, or do no more than apply theory to critique or prescribe practice. In *praxis,* neither theory nor practice serves as a foundation, as a validating ground for the other. Rather, *praxis* conjoins theory and practice—placing the two in dissonant tension.

Praxis demands interaction with audience. The traditional rhetorical paradigms view audience as the group of people a writer must persuade. The audience is perceived as the mostly passive receivers of the message (Porter, 1992). Even though workplace studies notice how meaning is negotiated socially among diverse writers, these studies tend to treat this negotiation not as between writer and audience but as negotiation among writers (i.e., as collaborative writing). They have persisted in maintaining the binary between author (or multiple corporate writers) and audience.

The central principle of social constructionist rhetoric is that meaning and knowledge are developed both historically and situationally through social interaction. The significant implication of this theory is that the audience-writer binary breaks down: Collaboration occurs between writer and audience, the audience becomes the writer, the writer becomes one with the audience. The ethical implication here then is not just that it is to the advantage of the writer to "analyze" audience, but that it is the writer's ethical obligation to "identify" with audience, negotiate meaning with the audience, and work to blur those roles that traditional rhetoric has staunchly maintained.

Whenever we write we enter into a cooperative arrangement with others. Any act of communication presupposes a set of agreements—an understanding about how we are to be with one another. Whenever we as writers use that set of agreements to manipulate our audience, we are writing unethically. That is a very easy claim to make because it begs the question (i.e., "manipulate" = "unethical"), and so no one could possibly disagree with it. And yet many of our most cherished ethical principles are similarly constructed. No one could possibly disagree with the universal principle "write clearly," because the difficulty is determining what *clear writing* means in any given situation, especially given the common problem of writing to multiple readers. Abstract principles do not alone provide us with answers to such questions. They can provide guidance perhaps—but decisions must be made locally, at the level of situation, considering the important contribution of audience.

The Role of Law and Policy
in Corporate Composing

Is it possible to construct a viable social workplace ethic based on critical *praxis*? What would such an ethic look like, and how would we go about constructing it? We can certainly call on the historical tradition of ethics and rhetoric as part of our critique, and we should certainly look to identification with audience (i.e., customers, clients, consumers). But we can also look to law and policy—which provide a very real documentary presence in the workplace—for additional guidelines.

Through unconscious practice or deliberate policy, corporations develop procedures for producing widely distributed documents such as product information, packaging, and advertising. These kinds of documents—for which the corporation serves as author—are common in the workplace. The corporate-level, mega-composing practices which produce such documents come under especially close scrutiny whenever a company is sued by customers or competitors. In recent research, I have observed how a lawsuit can influence both the reception and production of such documents (Porter, 1987, 1990a). In both cases a lawsuit made a company more aware of its composing practices and lead one company to review and revise those practices.

In one case (Porter, 1990a), several fertilizer companies were sued by a competitor for mislabeling their bags of composted manure. One of these companies countersued, on the basis that the plaintiff's product was similarly mislabeled. The product in question was a bag labeled "sheep manure" that contained only 7% sheep manure (and the rest other types of manure). The issue was whether this labeling constituted misleading packaging information: Was it unethical to label the bag in such a way if the quality of the product in no way depended on the percentage of sheep manure or if such labeling was a standard practice in the fertilizer industry?

How did the bag come to be labeled in such a manner in the first place? The problem may have been encouraged by a one-way, linear composing process (operations produces fertilizer, marketing takes responsibility for product packaging) that established clear lines of authority and responsibility to avoid interdepartmental political dissonance (all in the interests of efficiency). The lawsuit made at least one of the companies more aware of the interconnectedness between the product, its packaging, its consumers, and its competition; more conscious of the relationship between what they used to perceive as the manufacturing process and what we would view as the collaborative composing process; and more conscious that a little dissonance early can avoid problems later.

In another case (Porter, 1987), a medical insurance company was sued by a number of ex-policyholders for an alleged misleading advertisement. The advertisement included a reference to "lifetime protection"—which policyholders took to mean that the policy was "noncancelable." They were later unpleasantly surprised when their policies were canceled. The insurance company revised the ad, removing the reference to "lifetime protection," but not before these new (later canceled) policyholders were already enrolled. The problem may have resulted from an efficiency model that did not allow time for the policy writers to check whether advertisements met insurance industry guidelines. (Although, in fact, the plaintiffs accused the company of deliberately using the misleading advertisement to attract new policyholders quickly, in order to compensate for financial losses.)

In general, lawsuits can influence a company's view of composing in several ways:

- Litigation makes a company conscious that it *has* corporate composing practices, that these practices are ideological and political, and that inadequate composing practices can cost them a lot of money and get them in a lot of trouble. Litigation can motivate companies to review and critique their standard composing practices.
- Litigation makes a company realize that its documents are not merely vehicles for informing or persuading—but that they themselves are a kind of conduct—and, in the cases I have been examining, that behavior represents a relationship with the public.
- Litigation makes a company aware of the importance of critique. Litigation itself is a form of critique, built into the corporate system, that makes a company aware of itself in a more dramatic way (a more annoying if not permanently disabling way) than internal critique.

The main lesson that corporations can learn from these cases is that corporate composing models based solely on an efficiency principle can lead to problems, at least when they allow the principle of efficiency to override ethicality. When a large-scale corporate document moves from department to department within an organization, undergoing various phases of review, it is important not to neglect the kind of critique that might prevent such problems from occurring.

Corporations can implement several strategies to achieve ethical corporate composing, all having to do with project management of large-scale, corporate writing projects. Individually, writers can seek out alternative perspectives in reviewing documents, both in the design and invention phase of document development, as well as later in the process (Spilka, 1988, p. 220). Collaboratively, writers can make sure that the corporate composing process itself allows for diversity and critique throughout the document development process. Companies need an internal critique mechanism: a means by which corporate behavior can be challenged, questioned, and if necessary revised before problems result. Sometimes such a role is assigned to the legal department—but legal expertise is not the only authority or perspective that plays a role in determining the meaning of the developing document. This critique should not be exercised only at the end of the document cycle either. At that point, it becomes very difficult to undo something that has been done, even if it is clearly done wrong. Better to have the critique exercised early in the development cycle, during planning, design, and initial invention for a project. Large-scale documents—i.e., documents that play an important role in establishing the corporation's character or *ethos* (such as advertisements, brochures, quarterly reports, instructions, product packaging, warrants, and contracts)—should receive an ethical and legal check by more than one party. The ethical edit and the legal edit should be built into the composing process. Other tactics for ethical document development include bringing the audience into the design phase of document development (a strategy often practiced in usability testing of computer documentation) and including opportunities for interdepartmental consultation that will disrupt the possibility of linear and compartmentalized document production. For example, one company, motivated by the need for more effective product information, built into its product development cycle a procedure called *phase review*—a defined stage where a project team examines its own composing plan and steps outside the project to critique the process itself, to identify critical issues and problems and, if necessary, to alter the process (see Porter, 1990a).

This approach to document development is not necessarily more efficient: It requires time and effort and team participation. It is much easier for a single designated writer to apply a general ethical principle and make an individual decision. Checking *intertextual ethicality*—that is, reviewing the consistency and compliance of a given document in terms of laws, policies, and other corporate documents—requires doing research, conducting usability studies, negotiating competing principles, and so forth.

Is such an approach to ethics worth it to corporations? This question itself assumes the priority of an ethical philosophy that values efficiency and cost effectiveness above other values within the corporation. It extends from a general utilitarian ethic, but is actually a simplification of that ethic, resulting perhaps from a confusion between efficiency and effectiveness (i.e., what's fastest is not always what's most effective). What *praxis* says in response to this question is that corporations will have to make the decision of *worth* on a case-by-case basis. The lawsuits cited above recall moments of corporate crisis that came about because the efficiency principle failed to meet the complexity of a document's situation. The cases show that neglect of ethical and legal matters can lead to lawsuits, which may be many times more costly in the long run than the cost of increased attention to document development. We can also measure cost in terms of corporate image: What does it cost a company to lose the respect of some, maybe many, of its clients, through public embarrassment, or through earning a reputation for dishonesty, carelessness, neglect? Sometimes, however, documents are produced hastily not simply to save money in production but because a need must be met quickly. Haste can be in the service of the audience, too: a corporation genuinely wanting to meet an immediate need (as in the case of some computer documentation). Again, making an informed and intelligent decision requires balancing values, needs, and circumstances.

Finally what such an ethics requires is a pluralism. An ethical composing process requires input from a variety of competing perspectives, both from within and outside the corporation. It requires critique as well, and a legitimate interest in the opposing point of view, the position that says "you're wrong because. . . ." Corporations must learn to value dissonance.

Ethical and legal factors often take on the very concrete status as policy within government and corporations. In fact, policy writing is just now beginning to emerge as a distinct and important genre of professional writing (Porter, 1990c, 1991; Rogers & Swales, 1990; Speck, 1990). *Policy,* whether in the workplace or in government, refers to any statement that specifies norms or ideals of operation; to any statement that functions to set standards or criteria for behavior. Policy documents (e.g., employee handbooks) are becoming increasingly more important and more common in the workplace. They are a type of writing that many professional writers will need to learn to produce, as well as a type of document that will intertextually influence almost all other documents produced in the corporation. Professional writers need to be aware of the role of policy (as well as the role of law), but they need to understand the useful role critique plays in forming, changing, and opposing policy. Without critique, policy becomes static, insensitive to changing situations and constituencies. Policy and critique form a useful pair, together working in dissonant harmony: policy as the statement of ideals (the what-should-be), critique as the challenge to those ideals.

Here's an example, one common to many college campuses, that illustrates the complexity of policy: Students in a fraternity at Purdue University want to have a party—a simple and common enough desire, no doubt in the interests of the social good. Can this fraternity purchase alcohol for its party? To answer this question carelessly, without consulting the appropriate authorities, can lead to problems: Fraternity suspension is one possible outcome, but an intoxicated student being killed driving home would be a much more serious result. To answer this question, the fraternity needs to be aware of a number of interrelated, sometimes inconsistent policies and documents, as well as the general history of the problem. Federal law mandates that Purdue University must have a campus drug and alcohol policy. Purdue University (1990) does have such a policy, an interim policy, part of which refers to Indiana Code related to alcohol and drug offenses. In addition, two campus organizations—the Panhellenic Council and the Interfraternity Council—have adopted guidelines for fraternities and sororities on campus (Gerrety, 1991). They produce policy, in other words. But is this policy legally binding? What happens if a fraternity violates Panhellenic policy, but stays within Indiana law? Are members of a cooperative housing arrangement bound to the same policy?

Policy is by its very nature maddeningly ambiguous. One policy says that guests at a fraternity or sorority party may bring their own, privately purchased alcohol, but that the fraternity or sorority sponsoring the party may not collect funds to sponsor the purchase of alcohol. Does that mean that a fraternity brother can individually collect money from a few of his brothers to purchase beer to bring to the party? Would this fraternity member be violating either the letter or the spirit of the policy? (What exactly does *fraternity sponsored* mean?) What exactly is the binding status of the university policy? Is the university policy itself legal in restricting the use of alcohol by adults? The entire issue also has to be considered in light of historical factors: Attitudes toward excessive use of alcohol by fraternities and sororities seem to be in a process of dramatic change; the public is now less willing to overlook what was once considered simply standard college high jinks (in spite of whatever policy and law might have said). Lawsuits against fraternities and sororities found liable for irresponsibly serving alcohol have also influenced the climate in which these ethical decisions must be made.

Policy, like theory, functions as a guiding principle rather than as a determining rule in this realm. Just as the professional writer needs to interpret the injunction to write clearly for each specific composing event, the fraternity or sorority as a social agent needs to interpret and apply its own alcohol policy considering other perhaps inconsistent and ambiguous policies and laws, of perhaps uncertain or questionable jurisdiction, and in terms of specific behaviors. Here is the role of *phronesis*.

This example shows the complexity and intertextual nature of policy, and it shows how policy influences behavior. Another example will show policy as involved with composing practice. In 1990, the U.S. Congress passed a uniform food labeling bill (the Nutrition Labeling and Education Act of 1990) that provides guidelines and policies for food packaging and other types of food description (S. Pratt, 1990). Writing teams responsible for product packaging and marketing within the food industry will need to be aware of this legislation: They will need to know what products the bill does and does not cover, what types of information the bill mandates, and the constraints the bill places on language use in product packaging and advertising. (For instance, the bill will provide guidance with respect to "no cholesterol" claims.) They will also need to know when the bill passes into law and when their company will become responsible for meeting its guidelines. The bill itself does not provide absolute or comprehensive ethical authority, of course, but it does provide a very concrete (because textual) position, against which writers must review and critique their product packaging. Writers cannot afford to lean too heavily on the bill itself, however; they must know in what ways the consumers of their company's product may have distinctive needs, may require more (or a different type of) description. Ethical invention for such composing requires a comprehensive analysis of the situation, including review of U.S. Food and Drug Administration (FDA) regulations, industry standards and practices, consumer characteristics, and the particular marketing setting, as well as a consideration of general rhetorical and ethical theories.

The insurance company in the case discussed previously might well have avoided a lawsuit if it had consulted and heeded a variety of documents that would have provided practical guidelines, such as Federal Trade Commission (FTC) guidelines regarding fair advertising and, perhaps more important, insurance industry ethical guidelines (whether ratified or not). Insurance industry guidelines would have warned the writers about references to time in an insurance advertisement, particularly to noncancellability. The guidelines note that insurance advertising is different from other sorts of advertising, mainly because consumers tend to accept it more literally than they do beer or car commercials. Thus, although the FTC allows "puffery" in advertising generally, insurance advertising writers need to recognize how their particular genre requires distinctive treatments. They would be well advised to avoid terms like *lifetime*.

Conclusion

The ethic that I am developing here is not simply an idealistic or theoretical ethic. It refers to an abundance of codes, laws, and policies that exert an influence. The discussions of ethics that proceed by general theory alone do not go far enough to encounter the problematics of practice. General guidelines have to be interpreted, and this is where practical judgment of the writer comes in. She will also have to be aware of and responsive to contradictory policies and be able to critique and evaluate policies. As business becomes increasingly more litigious, the professional writer is going to need to be more watchful of the sometimes far-reaching policy implications of writing practice within the corporation. In fact, we can no longer teach professional writing strictly from the point of view of the corporation or the discipline. The significant parameters will be neither the discipline nor the corporation—but rather professional writers will need to be sensitive to a diverse network of concerns, extending among corporations, disciplines, and citizens.

On the other hand, I do not want to suggest that local law or public policy can provide comprehensive ethical guidelines. Actually, my point is that ethical practice cannot be determined by any single set of guidelines or theories. Ethical practice requires a pluralistic dialectic: The professional writer and the organization must locate a variety of ethical sources, available in theory as well as in policy and legal documents of various sorts, and carefully critique these in light of each composing event.

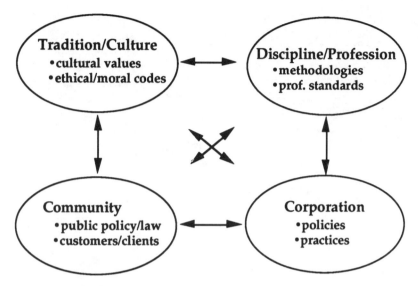

Figure 7.1. Writing Situation

Writers must carefully examine the situation of their writing practice. *Situation* is a fairly elastic term, of course. Figure 7.1 illustrates that situation in the broadest sense includes cultural and historical considerations, community parameters (both public and disciplinary), organizational constraints, as well as the demands of a particular composing act (e.g., audience and topic). It is with a sense of the constraints and guidelines of all these elements that the thoughtful and ethical writer composes. (Note, too, that Figure 7.1 does not provide a privileged standpoint for this critique; it only identifies some of the key elements of the dialectic.)

Philosophically, this chapter affirms a number of statements calling for a return to a humanistic basis for professional communication (Garver, 1985; C. R. Miller, 1979; Zappen, 1987, 1991). Such calls note the connections between the rhetorical and both the political and ethical, and note that an ethical rhetoric has as its proper aim the social good (however difficult it may be to agree on what that is). Note that this position differs from most conventional discussions of ethics, which have tended to focus not on the question of the social good so much as whether or not the individual is doing the right thing—that is, the focus has been personal salvation rather than social good.

I believe the humanistic perspective is vital to ethical corporate composing—at the same time I do not believe the humanistic tradition can critically justify claiming absolute priority over law, policy, or corporate composing practice. Again, we come to the issue of the theory-practice binary. Pointing to the humanistic tradition as providing the foundational grounds for evaluating corporate practice is another way of privileging academic/theoretical expertise. If we continue to do this, we can be assured of the continued irrelevancy of our humanistic perspective.

In practical and pluralist ethics, humanistic values are important because they serve heuristically to aid critique: They do not generate answers, but they are necessary elements of serious inquiry. And yet writers in the workplace need to look to other sources for ethical guidelines: law, policy, and professional standards. Ethical corporate composing requires an intertextual, pluralistic critique; it requires that corporations and writers investigate a variety of sources, both within and outside the corporate context, negotiating among them and weighing them against one another dialectically.

INTERPRETATION

8

Conflict in Collaborative Decision-Making

REBECCA E. BURNETT

Asking writers to collaborate is asking them to engage in a complex problem-solving, decision-making activity. In this chapter, I argue that this decision-making is more productive if co-authors engage in *substantive conflict,* which may be seen as a way to signal discrepant points of view. More specifically, I propose that substantive conflict is critical primarily because it serves to defer consensus; thus, collaborators have the opportunity to pose alternatives and voice explicit disagreements about both content and rhetorical elements.

In shaping this argument, I begin by discussing the way conflict is viewed in several disciplines and then briefly examine the evolution of collaboration in composition. In the bulk of the chapter, I present an observational study in the context of an upper-level writing class. Specif-

AUTHOR'S NOTE: David Wallace and Kathy Lampert were valuable supporters who provided plenty of substantive conflict as I worked through revisions of this chapter. Linda Flower, John R. Hayes, Lorraine Higgins, Juliet Langman, and Wayne Peck offered insightful observations and stimulating conversations at critical junctures along the way. This work was completed as part of the Making Thinking Visible Project at Carnegie Mellon, funded by the Howard Heinz Endowment of the Pittsburgh Foundation and sponsored by the Center for the Study of Writing at Berkeley and Carnegie Mellon.

ically, I analyze the decision-making of co-authors who use collaborative planning, and then I show the relationship between the kinds of decision-making they engage in and the quality of their documents. In the implications, I suggest fruitful areas for further inquiry both in the workplace and the classroom.

Conflict: Disciplinary Perceptions

While theorists, researchers, and practitioners in a number of disciplines encourage substantive conflict in their discussions of collaboration, their counterparts in rhetoric and composition have until recently urged consensus. The apparent differences stem partly from definitions and partly from the evolving understanding of what constitutes effective collaboration.

Defining Conflict

Substantive conflict can enhance collaborative decision-making, but two other kinds of conflict—*affective* and *procedural*—are not so valuable (Putnam, 1986). Affective conflict, which deals with interpersonal disagreements, is nearly always disruptive to collaborative decision-making. For example, when collaborators disagree because of personal prejudices (e.g., prejudices stemming from strong social, political, economic, racial, religious, ethnic, philosophical, or interpersonal biases), they are seldom able to focus on the task. Similarly, procedural conflict, which deals with disagreements about how the collaborators should work together, can also be disruptive if the problems are not resolved or managed effectively. Procedural conflicts can include disagreements about factors such as meeting dates and times, individual task assignments, group organization and leadership, and, curiously, methods of resolving disagreements. While unresolved procedural conflicts can prevent work on collaborative projects from even getting started, discussion of different procedural approaches can lead to a compromise that is mutually acceptable to the collaborators and productive for their decision-making.

While affective and procedural conflict can be detrimental to collaboration, substantive conflict, which focuses on alternatives and reaching *stasis,*[1] can be beneficial. Thus, collaborative writers need to discourage

affective and procedural conflict while at the same time encouraging
substantive conflict about content and rhetorical elements of the document they're planning. Substantive conflict during collaboration not
only is normal but can be productive (Galegher, Kraut, & Egido, 1990;
Johnson & Johnson, 1979, 1987; Putnam, 1986), in large part because
it gives collaborators more time to generate and critically examine alternatives and to voice disagreements on their way to making a decision.

Put another way, substantive conflict defers premature consensus (cf.
groupthink, Janis, 1982). Researchers and theorists in a number of disciplines—social psychology, decision theory, small group communication, cooperative learning, and computer-supported cooperative work—
argue that premature consensus can short-circuit effective decision-
making. Neglecting cooperative, substantive conflict can reduce the
effectiveness of a group and lower the quality of the group's decisions.
However, encouraging cooperative, substantive conflict can increase
the effectiveness of a group, improve the quality of the decisions, and
increase the group's commitment to the decisions that are reached (see
Putnam, 1986).

I propose that in collaborative decision-making about writing, substantive conflict includes two specific types of interaction about content
and rhetorical elements such as purpose, audience, conventions of
organization and support, and conventions of design. First, substantive
conflict includes considering alternatives—for example, when one collaborator suggests, "Let's do *x*," the other collaborator might respond,
"Yes, *x* is a possibility, but let's consider *y* as another way to solve the
problem." Second, substantive conflict involves voicing explicit disagreements—for example, when one collaborator suggests, "Let's do
z," the other collaborator might respond, "No" or "I disagree" or "I
think that's wrong." Both types of interaction have the advantage of
deferring consensus.

Evolving Views
of Collaboration in Writing

Collaborative writing practices are evolving—moving from a representation of collaboration that emphasizes construction and acceptance
of a social consensus as the primary goal to a representation that urges
engagement in explicit substantive conflict on the way to that consensus. The difference seems to be one of focus rather than fact, for discussions of collaboration typically include some reference to considering

alternatives (whether called brainstorming, debate, or dialectic) and voicing explicit disagreements (usually called devil's advocacy). The difference in the representations has to do with what happens on the way to consensus because, of course, the end result of a successful collaborative task is consensus.

Bruffee is most often associated with a position that encourages consensus. While he sees collaboration as providing a "social context for conversation" (Bruffee, 1984, p. 642), he also says that individuals "establish knowledge or justify belief collaboratively by challenging each other's biases and presuppositions" (p. 646). He acknowledges that teachers need to "help students negotiate among themselves to resolve differences of opinion and judgment, help them understand why such differences occur, and help them find information and gain experience that will enhance the quality of judgment finally arrived at" (Bruffee, 1985, p. 9). Although Bruffee clearly includes resolving differences as a part of collaborative interaction, he does not talk explicitly about alternatives or disagreements as critical parts of collaborative interaction that should be encouraged, even provoked. His emphasis seems to be on resolving differences as a means to reaching a larger goal, a socially constructed agreement, rather than encouraging alternatives and disagreements as a way to strengthen the eventual consensus.

Attitudes about the role of substantive conflict in collaborative writing, though, are evolving to incorporate views typical of those in other disciplines mentioned above. Simply, the value of encouraging substantive conflict is being recognized (Burnett, 1991). Some composition theorists and practitioners have argued that consensus in collaborative learning is "inherently dangerous . . . [because it] stifles individual voice and creativity, suppresses differences, and enforces conformity" (Trimbur, 1989, p. 602). The practice of encouraging consensus between collaborators in writing classes has started to give way to a view that encourages substantive conflict. For example, Karis (1989) builds an argument for conflict based on Burke, suggesting that "collaborators (especially students) need to be made more aware of the role and value of *substantive* conflict to the collaborative process" (p. 124).

Similar changes in awareness about the importance of conflict are evident in the arguments made by S. Clark and Ede (1990) who point out that Aronowitz and Giroux develop a notion of resistance that "helpfully complicates the way we may think about conflict in education" (p. 279). S. Clark and Ede (1990) argue that "resistance is not simply oppositional. Resistance opens up possibilities for learning for teachers and theorists, as well as for students" (p. 284). However, substantive conflict—a critical element in effective group process and decision-making—is still often ignored in classroom practices of collaborative writing. Specifically, classroom collaboration too often emphasizes consensus without what Trimbur calls "intellectual negotiation" (quoted in Wiener, 1986, p. 54) and ignores or suppresses conflict (cf. Ewald & MacCallum, 1990).

A Study: Substantive Conflict in Collaborative Decision-Making

The preceding discussion about evolving perceptions of substantive conflict draws attention to the complexity of collaborative decision-making. One question raised by this discussion is whether there is a relationship between the amount of substantive conflict in the decision-making of co-authors and the quality of the documents they create. To answer this question, I conducted a descriptive study in which I identified the substantive conflict in co-authors' decision-making, then evaluated the quality of their documents, and finally determined the relationship between these two factors.

Making Design Decisions

Four specific issues were important in the design of this study: determining the kind of collaboration to focus on, selecting the participants and the environment they would work in, creating a task and context for the task, and determining what kind of data to collect.

Focusing on Co-Authors

One of the first issues I needed to resolve was what kind of collaborative relationship to use as the focus for the study. I decided to concentrate on co-authors because they present special problems that foreground characteristics of collaboration often downplayed in other forms of collaboration. Co-authors are equally vested in the collaborative process and have an equal stake in the product; free riders are less likely. In the abstract, at least, this equal investment increases each collaborator's commitment to the task, participation in decision-making, and interest in the quality of the final document. The initial equality of co-authors in this study provided the opportunity for me to build in contrasting points of information that were intended as a foundation for legitimate differences in approach and interpretation in a way that reflects workplace realities.

Selecting the Participants

The second design issue centered on selecting the participants and the environment for the study. I decided to work with upper-level students in technical and business majors, most of whom were 2 to 6 months away from accepting a job in business or industry. These students would recognize the professional necessity of learning more about collaborative interaction, and, thus, take the task seriously. The participants were 48 students enrolled in three sections of an upper-level business communications course for industrial management and technical majors at Carnegie Mellon University. They were randomly assigned as pairs within each class section. The training that 24 pairs of students received about planning and collaboration as well as the task they completed were part of their regular course work; they were encouraged to recognize both individual and social factors that affected their writing and their collaborative efforts.

Creating the Task

The third design issue involved creating a workplace simulation, which included a writing task that was both topically and rhetorically complex. A simulation task can create a context that mimics the workplace, and, thus, provides reasons for students to engage in substantive conflict and make decisions. Not only can a simulation task encourage substantive conflict but it is also pedagogically relevant for students in an upper-level business communications course. Each of the 24 pairs planned and co-authored a recommendation report based on their collaborative analysis of an in-house company document and then recommended ways it could be revised as a product information sheet for customers in the company's expanding market.

The scenario allowed, even encouraged, substantive conflict in two ways. First, the complex technical content (solar heating system) and rhetorical elements were placed in a realistic context. This context included strong and sometimes conflicting management preferences, economic constraints in publishing the final document, designated areas of co-authors' expertise, expectations and needs of other departments, and political factors (e.g., the original in-house document was written by the company president). The students had to sort through complex content, resolve potentially conflicting information, and make decisions about multiple audiences, multiple purposes, genre considerations, and task constraints.

The second way that the simulation encouraged substantive conflict was giving each co-author in a pair a slightly different version of the scenario. They received different descriptions of a conversation they had individually with their immediate supervisor. These conversations provided the co-authors with unique information, appropriate for them to know because of their specializations. The differences provided a foundation for conflicts about approaches and interpretations in a way that reflected workplace realities.

Capturing Decision-Making

The fourth design issue focused on determining what data could best capture the co-authors' decision-making. Because I was investigating whether a relationship exists between the amount of substantive conflict in the decision-making of co-authors and the quality of their documents, I needed two specific kinds of data: an assessment of the quality of each recommendation report and a measure of the amount of substantive conflict co-authors engaged in during decision-making. To assess the quality of each recommendation report, I decided to evaluate holistically each document and then rank order them.

To determine the amount of substantive conflict co-authors engaged in during decision-making, I asked the co-authors to tape record their second collaborative planning session. Based on patterns of decision-making that had emerged in my pilot studies (Burnett, 1990), pairs were instructed to define and resolve any substantive conflicts and generate workable alternatives primarily during their second (of three) collaborative planning sessions. The tapes of these second sessions were transcribed and analyzed to identify the amount of substantive conflict in the co-authors' decision-making.

To help co-authors generate and focus their conversations about content and rhetorical elements, I asked them to use a heuristic called collaborative planning. This heuristic reminds writers to consider and reconsider content and rhetorical elements in ways that are typical of experienced writers but seldom considered by inexperienced writers (Flower et al., in press). Collaborative planning gives co-authors a forum in which they can engage in substantive conflict about content as well as rhetorical elements. The substantive conflict typically takes one of two forms, both of which defer consensus in decision making: indirect challenges that pose alternatives and direct challenges that voice explicit disagreements. Posing an alternative implicitly suggests that what is already being considered is in some way inadequate; the alternative may respond to that inadequacy. An explicit disagreement clearly signals dissatisfaction with what is being considered and often identifies that problem.

Analyzing the Data

I used two measures to see if the amount of substantive conflict in the co-authors' second planning session was related to document quality. First, I used a quality measure to evaluate the 24 co-authored documents, and then I developed a conflict measure to determine the percentage of substantive conflict in each pair's collaborative planning.

Selecting a Quality Measure

I holistically assessed the 24 recommendation reports for quality and then rank ordered them, considering content and the following rhetorical elements: purpose, audience, content, organization, support, and design. A second expert writer/teacher independently assessed the same reports. We had a very high level of agreement on the quality of the reports; a Spearman rank-order correlation (ρ) on our holistic assessments of the documents' quality was +0.943.

Designing a Conflict Measure

I developed a measure that identified four kinds of decision-making to distinguish substantive conflict from other kinds of decision-making. I focused only on episodes in the second collaborative planning sessions that dealt with content or rhetorical elements; episodes that dealt with affective or procedural topics were eliminated from the analysis. Assessing the amount of substantive conflict required two steps: identifying topical episodes in the transcripts of the second collaborative planning sessions and coding these episodes for types of decision-making.

In the first step, I identified topical episodes, which are chunks of coherent conversation that have definable topical boundaries. Using topical episodes for analysis—rather than, for example, idea units or conversational turns—enabled me to examine chunks of collaborative interaction that led to decisions. Episode boundaries are typically identified by a topical shift or change in the rhetorical focus, by a decision or an agreement to defer a decision, or by an unannounced shift to another topic.

In the second step, I coded these topical episodes to identify four categories of collaborative decision-making: *immediate agreement, elaborating a single point, considering alternatives,* and *voicing explicit disagreements.*

I defined the combination of two categories of decision-making—considering alternatives and voicing explicit disagreements—as substantive conflict because both are ways to challenge an existing idea. Considering alternatives gives collaborators the opportunity to identify and evaluate other possibilities, thus enlarging the number of choices without ever saying to the other person, "I don't like your idea." Instead, a collaborator can say, "Another way to look at this might be. . . . " Considering alternatives also forces collaborators to identify criteria for decision-making because once two or more possibilities exist, collaborators need some way to select the best one. While considering alternatives indirectly challenges an option, voicing an explicit disagreement directly challenges. Such clear signals of disagreement identify a collaborator's position and often open the door for posing alternatives that respond to the concern. Considering alternatives and voicing explicit disagreements defer consensus by requiring collaborators to do more than simply extend an existing point.

Elaborations of a single point also defer consensus in decision-making; however, collaborators typically maintain a consistent point of view and merely provide details about a point that they both agree on. The benefit of such elaboration is that collaborators add examples and sometimes even explanations for a particular point. While this is nearly always necessary, elaborating a single point is insufficient if it's the only way that collaborators defer consensus because it presumes the idea being elaborated is the best option. The problem isn't with elaboration but with elaboration to the exclusion of other forms of deferring consensus.

The three categories of decision-making described so far—elaborating a single point, considering alternatives, and voicing explicit disagreements—are all kinds of deferred consensus. They enable co-authors to put off reaching consensus. All three kinds can be valuable, but their value is influenced by the relevance of the content being considered. If co-authors chose to elaborate a point, explore alternatives, or voice disagreements about topics that are silly or irrelevant, they succeed in deferring consensus, but their interaction would be unlikely to improve their decision-making.

The final category of decision-making is immediate agreement (i.e., consensus), a behavior that is essential in all decision-making. Immediate agreements are sometimes embedded in other kinds of decision-making; they are important conversational turns that let co-authors know if their partner agrees with the direction of an elaboration or argument or understands a disagreement. Immediate agreements are only detrimental if they are the predominant kind of decision-making.

Using the four categories of decision-making, I coded the topical episodes in the 24 transcripts of the second collaborative planning sessions. A second rater coded 20% of the 24 collaborative planning sessions, a total of 5 complete transcripts. Because the sessions varied tremendously in length, the second rater coded a random sample stratified by length. The second rater's agreement with my assessment of the types of decision-making was 0.89. The coding allowed me to identify different kinds of decision-making and then determine the relationship between substantive conflict and the quality of documents that co-authors produced.

Establishing the Relationship Between Document Quality and Substantive Conflict

Was there a relationship between document quality and substantive conflict? The simple answer is yes. The first part of this section focuses on the quantitative results of the study; the second part examines examples of co-authors' decision-making.

Reviewing the Results

To determine if a relationship existed between document quality and co-authors' percentage of substantive conflict in decision-making, I first

separated the documents into two groups according to document quality: High-quality documents were the reports ranked 1 to 11.5 (2 reports
tied for ranks 8 and 9; 2 others tied for ranks 11 and 12), and the
low-quality documents were the reports ranked 13.5 to 24 (2 reports
tied for ranks 13 and 14; 2 others tied for ranks 20 and 21). Table 8.1
shows the 24 pairs of co-authors ranked according to the quality of their
documents and separated into high-quality and low-quality documents.
Columns 1 and 2 list the rank order of the 24 documents and the co-
authors who wrote those documents. Columns 3 and 4 show the percentage of episodes in which alternatives were considered and the percentage of explicit disagreement—the two categories that comprise
substantive conflict. Column 5 shows the total percentage of substantive conflict each pair of co-authors engaged in during their collaborative decision-making. Notice that the pair of co-authors who wrote the
top-ranked document spent nearly half of their topical episodes engaged
in substantive conflict, whereas co-authors of the bottom-ranked document spent very little time engaged in substantive conflict. Overall, the
co-authors of high-quality documents spent an average of 30.2% of
their episodes in the second collaborative planning session engaged in
substantive conflict. In contrast, the co-authors of low-quality documents
spent an average of 10.8% of their episodes engaged in substantive
conflict.

I also separated the co-authors into two groups according to the percentage of substantive conflict in each pair of co-authors' decision-
making: High-conflict sessions contained above-the-median percentage of
substantive conflict; low-conflict sessions contained below-the-median
percentage of substantive conflict.

Figure 8.1 shows the pairs of co-authors who had high/low *substantive conflict* (i.e., the combination of considering alternatives and voicing explicit conflict) and those who produced documents of high/low
quality. Collaborators who produced high-quality documents typically
engaged in substantive conflict; that is, they considered more alternatives and voiced more explicit disagreements than collaborators who considered few or no alternatives and voiced little or no explicit disagreements. The statistically significant χ^2 value of 10.66 ($df = 1$; $p < .005$)
shows that there is a relationship between document quality and the
substantive conflict that co-authors engaged in. There is a 0.5% chance
that the relationship between the quality of students' documents and the
amount of conflict is accidental.

Table 8.1   Pairs of Co-Authors Ranked According to the Quality of Their
                 Documents

Quality Ranking of Co-Authored Documents		Considering Alternatives (Percentage)	Voicing Explicit Disagreement (Percentage)	Total Percentage of Substantive Conflict
High-Quality Documents				
1	Rick — Maggie	36	13	49
2	Kevin — Neal	19	2	21
3	Matt — Dorothea	14	11	25
4	Ed — Anna	22	10	32
5	Aaron — Yang	35	4	39
6	Betsy — Azeeta	24	1	25
7	Mason — Greg	29	7	36
8.5	Jesse — Inga	32	10	43
8.5	Sean — Ben	25	0	25
10	Joyce — Ling	21	5	25
11.5	Frank — Lou	13	1	13
11.5	Adam — Ray	23	6	29
	Mean	**24.4**	**5.8**	**30.2**
Low-Quality Documents				
13.5	David — Ming	5	0	5
13.5	Rich — Caren	10	0	10
15	Carl — Drew	0	0	0
16	Sam — Mitch	21	0	21
17	Julie — Mel	15	0	15
18	Dean — Sujit	12	0	12
19	Justin — Ted	2	0	2
20.5	Linda — Chung	12	0	12
20.5	Jessica — Margaret	10	0	10
22	Anthony — Li	14	2	16
23	Rena — Marc	20	4	24
24	Josh — Pete	3	0	3
	Mean	**10.3**	**0.50**	**10.8**

NOTE: The table also includes co-authors' percentages of considering alternatives and voicing explicit disagreements as well as the total percentage of substantive conflict as part of the decision-making for each pair.

Examining Co-Authors' Decision-Making

Understanding the relationship between substantive conflict and document quality requires examining all four kinds of decision-making because substantive conflict doesn't occur in isolation; instead, it occurs in the context of other kinds of decision-making. The examples in

	High-quality Documents	Low-quality Documents
High Conflict	10	2
Low Conflict	2	10

Figure 8.1. The Relationship Between the Quality of Documents and Collaborators' Decisions Based on Substantive Conflict (i.e., alternatives and explicit disagreements)

this section illustrate the four categories of decision making in this study and raise questions about the results of this study.

Immediate agreements are an important kind of decision making that occur in virtually every kind of collaborative interaction. Immediate agreements are conversational moves that signal active listening and allow the interaction to move forward. Although the paraphrases, repetitions, and agreements that are typical of immediate agreements add no new content to the conversation, immediate agreements are an essential part of any collaborative interaction, keeping the conversation going by acting as backchanneling cues. The following example shows co-authors Sean and Ben reaching an immediate agreement during their discussion of the document they're analyzing:

Sean: I can't believe they're sending this to customers.
Ben: I know. It's terrible.

All co-authors use immediate agreements similar to Sean and Ben's, but co-authors who produced low-quality documents in this study used them twice as frequently as co-authors who produced high-quality documents. Co-authors who produced low-quality documents sometimes seemed to use immediate agreements rather than deferring consensus, which would have enabled them to elaborate points or engage in substantive conflict. Immediate agreements should not replace careful examination of content or rhetorical elements. In this study, co-authors who produced low-quality documents often agreed immediately without hearing any rationale or explanation for a statement.

Even when co-authors agree immediately, they should consider occasionally reexamining their consensus. Deferring agreements has several benefits, the most important of which are increasing the effectiveness of a group, improving the quality of their decisions, and increasing their commitment to the decisions they do reach (Gouran, 1986). But deferring agreement is not, by itself, of any particular value: What happens during the time before consensus is reached is critical.

The first type of deferred agreement, elaborating a single point, provides details and explanations. In the following brief example, Sean and Ben discuss the trim size of the document they're planning:

Sean: It could be more like, for size—
Ben: 8½ × 14.
Sean: Right. It could be more like that and all you have to do, instead of folding it three ways, you could fold it in half and then fold it again if you're using a legal-size envelope.
Ben: Yeah.
Sean: As long as it's still just 8½ inches wide, it fits in any normal envelope.
Ben: That's true. Maybe we should go with that.

Sean and Ben are refining and clarifying their point about document size—a point that helps them move toward decisions about the document they're planning.

In this study, elaborations of a single point such as Sean and Ben's were the single largest category of topical episodes, comprising more than 75% of episodes of co-authors who produced low-quality documents and nearly 66% of episodes in co-authors who produced high-quality documents. Simply having episodes that elaborated single points did not translate to high-quality documents.

What led to this Sean and Ben's episode, however, was the previous episode—a way to defer consensus that involves *considering an alternative,* a kind of substantive conflict:

Ben:	Hey, by the way, nobody said the information sheet has to be $8\frac{1}{2} \times 11$ paper.
Sean:	Yeah. It could be bigger paper. I've seen information sheets that were bigger paper.
Ben:	Yeah.
Sean:	You can still fold them up and put them in a legal-size envelope.
Ben:	That's true.

In this study, deferring consensus by considering alternatives was the second most frequent kind of topical episode. However, the difference in the percentage of alternatives considered is worth noting. Considering alternatives occurred almost 2.5 times more in the episodes of co-authors who produced high-quality documents as those who produced low-quality documents. Specifically, co-authors producing low-quality documents considered alternatives in approximately 10% of their episodes, whereas co-authors who produced high-quality documents considered alternatives in almost 25% of their episodes.

Another way to defer consensus with substantive conflict involves voicing explicit disagreement. An example of explicit disagreement comes from the collaborative planning session of yet another pair of co-authors, Ed and Anna. They are discussing whether the cost of the solar system should be included in the product information sheet. Ed has maintained that cost shouldn't be included; Anna disagrees.

Ed:	I don't think it [the cost] should go in there, because it's so different between people. Like they see like—a system that costs them 8,000, all of a sudden we quote them some price of 20,000. I mean—I just—I think we'd be better off keeping off, keeping out of there.
Anna:	Okay, but they'll want to know.
Ed:	Well, I mean, I don't know, if you mean—We should, we should talk about it if you disagree.
Anna:	I kind of do, because, um, I mean if somebody says, "Can I have your information sheet," they're going to want to know about

the price. They're going to want to have the price in front of them, especially if they're going to compare it to other systems. They're going to want to see the, the price, how much it's going to cost. Like let's face it, everybody's concerned about price. And I guess the sales people could have a different, a separate price sheet.

Ed and Anna manage their explicit disagreement by exploring reasons for it, which leads them to consider additional alternatives. These alternatives give them specific content for the document they're planning.

Ed: We, we, I mean, I would say I probably consider it's in a low cost, it says, "Its low cost should convince you to install a Sundance system in your home or business." Um—

Anna: Even so, the range from 4 to 26 [\$4,000 to \$26,000] would at least give them a ballpark figure. Or we could find out a little bit more detail on the cost.

Ed: Well, I thought we should put in a section like, depending on how big or how many you get, and just put all these things in there. Maybe we could put it in the advantages thing, if it is a low cost.

Anna: Okay. Well, that's true. We could stick it in. Maybe we could stick it in and say, "Cost is one of the advantages" and put the price in there.

As Ed and Anna consider alternatives, they often elaborate a variety of points and frequently express immediate agreements. Their explicit disagreement followed by consideration of alternatives is a pattern typical of co-authors who produced high-quality documents.

This pattern of disagreements followed by alternatives resulted in longer planning sessions for co-authors who produced high-quality documents than for co-authors who produced low-quality documents. Engaging in substantive conflict meant that some co-authors took more time to complete the task; however, this increased time alone is not a likely explanation for differences in quality.

Although the coding in this study was not sensitive enough to distinguish between various types of alternatives and disagreements, the substance of the substantive conflict is critical. For example, in their conflict, Ed and Anna come up with a number of ideas that influence the content and design of their report. In contrast, the only substantive conflict of Josh and Pete (co-authors of the lowest-ranked report) involves an argument about which font to use for their report. It is one of their few extended considerations of any topic; however, even if they continued this conflict for several more minutes, it would probably have made little difference in the quality of their report. Rather than offering explanations or rationales for one font instead of another, their conflict was little more than "I like Geneva best" followed by "I don't like Geneva. I like Chicago" followed by "Well, I really like Geneva." The conflict was settled not by examining possible reader reactions to those fonts but by the partner responsible for final report copy. Productive substantive conflict involves more than time; co-authors should deal in a serious way with topics of substance. Co-authors who produced high-quality documents posed alternatives and disagreed with each other, but they also offered justifications and explanations, considered opposing views, and tried to create sound arguments. They deferred consensus for a purpose: to make better decisions.

In this study, the two ways of deferring consensus through substantive conflict—considering alternatives and voicing explicit disagreement—were nearly always part of the decision-making of co-authors who produced high-quality documents. In contrast, both types of substantive conflict were far less frequent in the episodes of co-authors who produced low-quality documents. Deferring consensus through substantive conflict gave co-authors the opportunity to improve the quality of their decision-making. For example, they could develop rationales for their ideas, identify the strengths and weaknesses of their individual and collaborative positions, and pose more effective arguments.

Implications

The results discussed in this chapter demonstrate a statistically significant relationship between the type of decision-making co-authors used and the quality of the product they produced, a correlation that writers need to know about. Simply put, the co-authors who engaged in a high percentage of substantive conflict were clearly related to those who produced higher quality documents. These results suggest that classroom and workplace co-authors should consider the potential value of engaging in substantive conflict as they collaboratively plan documents.

Confirmation of these results in further studies could have important implications for both pedagogy and practice. Co-authors and writing teams in the classroom and the workplace could focus on the *process* of collaboration, recognizing that the nature of their interaction and decision-making could influence the quality of the document they create. Further investigations could also look for other possible explanations for the results reported here. For example, the category of deferring consensus by elaborating a single point needs investigation. This large category could be refined to identify distinct kinds of elaboration, ones that may reveal important relationships between the kinds of decision-making and the quality of documents. Another study could investigate whether there are factors that can predict document quality (e.g, substantive conflict, ability of co-authors, partner dominance).

Further studies might also explore topics related to areas of conflict and collaboration only hinted at in this chapter. One area for investigation could probe the relationships among substantive conflict, affective conflict, and procedural conflict. While substantive conflict is undeniably valuable in this study, it may never get a chance to help decision-making if the collaborators are embroiled in destructive affective and/or procedural conflicts, so learning about all kinds of conflict and the relationship among them is important. Another area for investigation could explore ways that co-authors manage substantive conflict: how they resolve discrepant information, what they do with alternatives, and how they deal with disagreements. More specifically, how do collaborators in different kinds of relationships deal with conflict? For example, if one of the collaborators is dominant, is substantive conflict possible? An additional area for investigation could examine collaborative planning as scaffolding that can help inexperienced writers become better as they provide support for each other. Collaboration in general and collaborative planning in particular are based largely on the idea that working together may be more productive than working individually. Building on Vygotsky's (1986) notion of a "zone of proximal development," Bruner (1978) saw scaffolding as a particular kind of "working together" in which a supporter encourages, guides, or assists a less-experienced or less-skilled collaborator to complete a task that could not be done individually.

Note

1. Dieter (1950) defines *stasis* as the "standing still, which must necessarily occur momentarily in-between opposite 'charges' and in-between contrary motions" (p. 369). In other words, stasis is the stopping place at which people can recognize and identify the points that are in opposition, those issues that are contested. Acknowledging and defining an issue are essential. Without an identified point of contention, specific disagreement is unfocused and alternatives cannot be posed.

INTERPRETATION

9

Validity and Reliability as Social Constructions

JANICE M. LAUER

PATRICIA SULLIVAN

This chapter argues that reliability and validity are socially constructed criteria that have been developed by the empirical community to guide its interpretive acts. These criteria serve as ground rules for discriminating the quality, value, and credibility of studies and results. The two notions are not transcendent, self-evident characteristics inherent in the nature of empirical research, but rather rules of evidence and inference agreed on and continually modified and refined by the research community.

Because these criteria are imbricated in the history of both qualitative and quantitative research as interpretive enterprises, they partake of the social nature of empirical research itself. Counter to charges of objectivity, the activity of empirical research is never disinterested, but always driven by many cultural factors. The problems studied are usually those deemed possible and worth investigating by the research community, which is situated in particular economic and political contexts in the academy and the larger culture. If researchers want to study other problems, to resist, their new problems are still related to the community, triggered by anomalies festering in its intertext.

In addition, researchers' knowledge constructions are mediated by many other factors—prevailing theories of writing, language, and knowledge; current discursive practices in the research community; the economic and political status of scholars; the perceived value of the research for creating cultural capital for users; and so forth. Research designs also constrain and shape researchers' acts. Bourdieu (1977) characterizes qualitative research as *subjectivist:* The ethnomethodologist has "presuppositions inherent in the position as outside observer, who in his preoccupation with *interpreting* practices, is inclined to introduce into the object the principles of his relation to the object," (p. 2). He similarly describes all empirical research as partial representations or perspectives that are adequations but never equal to the primary experience. In this intersubjective research world, validity and reliability authorize the possibility of constructing systematic knowledge in the face of randomness. They are a way the research community participates in the making of knowledge.

In this chapter, we discuss validity and reliability as social constructions and then demonstrate how the treatments of validity and reliability in three studies in professional writing functioned (or could function) as socially constructed arguments.

Validity and Reliability

Validity

Validity signifies that a piece of research and its claims are compatible with the community's theoretical structures, assumptions, and paradigms. Validity claims that what was observed can be interpreted inside the community's interests. When researchers argue for validity, they try to establish consensus with the rest of the discourse community, not about observations as in reliability, but about governing theories. The research community expects studies to help solve its salient problems, working within its conceptual frameworks. If researchers can demonstrate that both their problems and results are compatible with an acceptable theory, this argument bolsters their claims. If investigators go beyond these conceptual boundaries, they must first construct and argue for new theories and then for specific results. Either way, validity plays an important role in the community's acceptance of the results as reasonable and valuable explanations of the experience studied.

Because design practices—from ethnography to metaanalysis—are also shaped and changed by prevailing conceptions of the relationship among discourse, knowledge, and experience, they help establish validity. Van Maanen (1988) comments on this point:

> It is relatively easy to see that during particular periods (and within particular theoretical circles) ethnographers were able to more or less agree on what kinds of cultural interpretations were acceptable and authoritative for the ordinary purposes. Yet these largely implicit agreements were hardly fixed or timeless, since they eventually broke down as new ways of handling previously unseen representational difficulties emerged. (p. xi)

Researchers can establish validity by a variety of other cooperative means. They can argue that their constructs meet communal expectations by showing that the subjects in their studies demonstrate the individual differences expected by the community, e.g., differences in social class, maturation, or education. If constructs do not violate those expectations, researchers have an argument for validity. Investigators can also argue for validity by demonstrating that their measures forecast others' judgments or that their results concur with those in other studies. They can also engage experts to attest that topics or items in their survey-questionnaire, for example, get at the construct intended (Anderson, 1985; Lauer & Asher, 1988).

In addition, arguments for validity betray a study's ideology, especially in the literature review, which displays a researcher's theoretical allegiances. In the second part of this chapter, when we examine validity and reliability as socially constructed arguments in three professional writing studies, we also illustrate how these arguments reveal the studies' underlying ideologies and engage the researchers in invoking different intertexts. We use Faigley's (1985) categories of ideology (textual, individual, social) to characterize these allegiances. A *textual* ideology assumes the preeminent importance of finished discourse, directing researchers to analyze such features as syntactic characteristics, vocabularies, and genres of professional writing and to cite work on discourse analysis, research on sentence-combining, studies on readability, and so forth. An *individual* ideology values and directs researchers to study cognitive development, composing processes, and writing strategies and to cite work on these subjects. A *social* ideology valorizes social context as an object of study—social roles, group purposes, communal organization, and theories of culture—and engages a researcher in citing cultural studies, theories of discourse community and social construction of knowledge, and so forth.

All of these validity arguments work together with reliability proofs to socially construct new interpretations of observed experience.

Reliability

Reliability argues that what was observed by the researcher can be viewed similarly by others. It insists that claims about patterns and structures of experience can only be made within a discourse community, which expects multiple observers, observations, or points of view to be included in the analysis. No results are credible if they represent solitary effort. Thus reliability guards against idiosyncratic interpretations made on the basis of isolated observation.

Researchers are guided by socially established means of data collection and by expectations for triangulation of observation or acceptable statistical levels of significance and effect sizes. To meet these expectations, researchers must provide fine-grained, richly specified, and/or statistically calculated accounts of the patterns and interpretations they are advancing to enable others to co-create them. Goetz and LeCompte (1984) divide the ways in which credibility is established for qualitative studies into what they call internal reliability (that others can or did use the construct in the same way as the researcher) and external reliability (that others would uncover similar constructs in similar situations). Lincoln and Guba (1985) point out that such *dependability* is vital to naturalistic inquiry. A similar criterion operates in other scholarship such as historical and hermeneutical studies in which interpretations are carefully elaborated so that the community can understand, assess, and it is hoped, adopt them as preferable ways of symbolizing the social situation or the textual patterns under scrutiny.

Empirical researchers use reliability checks both during the process of analysis and in the publication of results. Early tests for agreement function heuristically, because if investigators have trouble enabling others to create the patterns and constructs they have developed, they can either adjust or abandon their interpretations or increase the specificity of their definitions or descriptions. Later, in the published account, researchers use reliability as one of their arguments within the community for the strength of the interpretation.

We examine below the role of these criteria in three professional writing studies, illustrating the ways in which validity and reliability have been or could have been used as communal warrants for the studies' claims. Using Faigley's (1985) taxonomy of ideologies discussed above, we point out some ways in which validity arguments articulate a study's ideology, either explicitly or implicitly.

Validity and Reliability in Three Studies of Professional Writing

Study 1: "Interactive Writing on the Job"

In Couture and Rymer's (1989) study, "Interactive Writing on the Job," validity arguments show that the study situates itself within two problem areas already deemed important by the discourse community: writing in the workplace and collaborative writing. The researchers, therefore, are expanding the understanding of such writing, not arguing for context or collaboration as valuable sites of research. The investigators assume that their results, if illuminating, will be welcomed by the community. They also count on the acceptability of their survey instrument design within the parameters of both composition inquiry and professional writing research. Their validity arguments draw on all three of Faigley's ideologies—social, individual, and textual.

A social ideology is explicitly invoked. At the beginning of their account, Couture and Rymer cite research on collaborative writing in the workplace, studies by Faigley and Miller; Lunsford and Ede; and Paradis, Dobrin, and Miller. They assume that a study of professional writers is consonant with the socially valued practice of studying writing in its actual contexts. While they also count on the acceptability of a survey design, they could note its limitations: that it cannot provide a rich, in-depth account of the social and cultural context. They do acknowledge that their findings about the world of work should not drive the classroom because many business practices are not worthy of imitation.

Within the purview of a social ideology, they begin to examine the economic, social, and political status of the writers by trying to determine their motivation for writing. The survey reveals that both career writers and professionals think that most of their work is assigned but that professionals believe they also write on their own initiative. Couture and Rymer (1989) point out that this perception may be due to the fact that such writing is difficult to distinguish from assigned writing because "routine documents are typically part of larger assigned projects; when viewed in isolation, however, they may seem to be done on the writer's own initiative . . . but the typical documents of the professional who writes—are part of the writer's overall responsibility for some larger objective. A management-set goal often governs the shape or directions of such documents" (pp. 81-82). The implications of these conclusions could be further probed, indicating that these writers probably occupy the lower economic and social ranks in their institutions and typically play political roles as perpetuators of the institutional goals rather than resistors. Couture and Rymer (1989) hint at this implication when they call for studies of how "preparing team documents may affect a writer's sense of professional ethos, revealing, for instance, whether a writer can represent himself or herself with integrity as a professional while adequately accommodating the stance of the group" (p. 88).

The individual and textual ideologies more tacitly penetrate the study. The researchers organize their survey of writing practices in the workplace around the composing processes: planning, drafting, revising, and objectives. They assume widespread understanding of these composing acts but they do not cite composing process theorists. Their findings are presented around the relationship between collaboration and the composing process:

- Writers interact during the composing process, both before and after drafting.
- Writers interact with those who have a stake in the written product and some vested interest in it.
- Writers get feedback on their documents before sending them out, but usually take responsibility for revising their own texts. (Couture & Rymer, 1989, p. 89)

The section on implications for further study calls for research on the "frequency of interaction throughout the composing processes—during inventing, planning, organizing, drafting, revising, and editing" (p. 88) and suggests the value of studies of novice/expert interactions. In addition, Couture and Rymer indicate that they studied collaborative processes across types of textual genre: routine writing (memos, letters, and progress reports) and special writing. Thus their study is ideologically mixed—

with explicit allegiance to the social, but also deploying the individual and textual.

Couture and Rymer rely less on reliability arguments. To meet the research community's standards, they report that they collected data from enough respondents to make comparisons and that the differences between professionals and career writers were significant at an acceptable level. A number of other reliability arguments could be used to bolster the social construction of the study. The researchers could (1) establish stability over time by repeating the questionnaire at a later date, (2) identify the total population from which the sample was taken, (3) indicate whether the sample was randomly selected, (4) mention the percentage of response received, (5) establish that subsets of items on their questionnaire had reliable agreement with other questions or with items on another survey, or (6) report the level of agreement that the Professional Writing Panel reached about its interpretations. Because all of these reliability arguments meet the expectations of an empirical research community, they could increase the persuasiveness of the study.

Study 2: "Process and Genre"

In "Process and Genre," Eiler (1989) develops validity arguments to accompany the simultaneous study of composing processes, genre, classroom writing, and workplace writing. She brings together a number of problem areas that are recognized among those in the research community but not normally conjoined: process and text, classroom and workplace. She also deploys several methodologies—experiment, case, and textual analysis—in service of her complex goals, namely to demonstrate that "the act of writing should not be divorced from its goal—the type of discourse the writer is attempting to create" (Eiler, 1989, p. 43). By using three separate studies to argue the points and by positioning her article in a way that brings together research focuses (process and product) and sites (academy and workplace) that are not normally considered together, the researcher captures questions of interest to the professional writing community but maneuvers in an area of ideological debate. She might assume that most researchers will be interested in her findings, but some may not readily accept the claim of importance of product to process; she is arguing for text as insightful to process. Similarly, Eiler might count on her experimental design gaining acceptability from the empirical research community and her textual analysis gaining acceptability from the linguistic research community. To the extent that she is addressing process researchers who may not value textual insights, Eiler must argue that textual studies shed light on important process issues, a point she argues through the pedagogy she suggests in the end of the article.

Eiler's (1989) studies display, on the balance, a textual ideology. In her initial arguments aimed at establishing validity, Eiler cites Halliday from linguistics and Kinneavy from rhetoric to discuss genre from a textual features perspective. Later, as she presents the three studies, the researcher bolsters her analysis with discussions of text linguists such as Hassan, Halliday, Fowler, and Lehrberger and Kittredge. Although her problem statement includes cognitive issues, she does not reference a cognitive theorist or explore the problem from the standpoint of individual composing choices. She also does not frame the study in terms of social ideology. Such a view might focus on the contrast between the world of classroom and the world of work, exposing the connections between the power positions of the writers and the types of responses they generate; might examine the collaborative roles of the writers in the second study; or might focus on the power relations of the lecturer and the writer/editor in the third study. The theoretical and the methodological company she keeps is in text linguistics. True, the article is not totally textual. Eiler's title and opening argument focus on the intersection of composing processes and genre, and the first study discussed involves an examination of composing choices, but the researcher resists the individual perspective. Though she is studying composing processes, particularly in her first study, the researcher's goal is to demonstrate the importance of genre to process. Then she reinforces the textual position by treating three studies as cases that progressively display textual ideology, relying more frequently on textual evidence and linguistic argument, moving away from questions of individual composing and ever more closely to describing and examining linguistic features vital to a textual view of the problem.

Because Eiler's methods differ across the studies, she deploys differing specific discussions of reliability. The first study's method is governed by the requirements of studies in the individual perspective and thus is concerned with assuring the readers in that discourse community that the conclusions drawn about writing processes are reliably reached. Operating under the aegis of experimental design, an important part of her case includes demonstrating the reliability of the sample chosen, the quality of the writing, and the students' grouping into levels of maturation to assure readers that the groups contrasted were actually separate groups. Eiler is particularly careful in relating how the 15 students who participated were selected (sampling), with the agreement of four teachers' judgments on the quality of their writing an important point of discussion (interrater reliability). She builds a case that measures maturation, so is careful to show that all the students were comparable, good students, whose writing quality groupings had the backing of four teachers' judgments.

In the second and third studies, however, Eiler shifts from an experimental design to a case design, deploying the case as a text linguist would, to demonstrate or illustrate. The second study narrates the interaction of two writers writing a book for physicians on the subject of demographics. At the start of the discussion, Eiler does not explain her methodology as she did in the first study, nor does she explain the study as a case study or ethnography. Instead, she calls on Hasan's analysis of the structure of nursery tales, invoking methods from text linguistics for this case and dispensing with readers' expectations that she will uphold her methodology to inspection for empirical reliability checks. The third study (which was published in its entirety in Couture, 1986) presents an abbreviated discussion on the translation of an oral lecture into a textbook chapter on physics. It uses linguistic methods to analyze "sentence openings for linguistic features like nominal and pronominal subjects, nominalizations, coordinators, conjuncts, disjuncts, and other categories occurring in fronted positions in the English sentence" (Eiler, 1989, pp. 58-59). Again, textual questions and methods drive the discussion of this case, and reliability is handled indirectly through the credibility of the examples. Had Eiler been invoking the reliability arguments for empirical cases in her second or third study, she might have discussed the selection of subjects, described the sites in detail, discussed methods for gathering data from a variety of sources or observers, or used several raters' judgments about the textual decisions. Her direct discussion of reliability in the first study reinforces a view that she saw the second and third cases primarily as cases of textual analysis. If the researcher had viewed the cases as studies of individual (composing) processes, she might have charted the activities and composing choices made by the writers involved in the case studies. If she viewed the cases as studies of social processes, she might have examined the political or social dimensions involved in shaping the American Medical Association (AMA) book project or the social contexts of genre, thereby problematizing the ideal notion of genre that the authors build in that book project. Process researchers in particular would have welcomed detailed discussion of process in the second and third studies; social researchers would have welcomed more detailed discussion of relationships to context.

In her discussion of the joint findings in the three studies, Eiler returns to the tension between process and product, using her findings for the importance of genre/goal to call for an instructional paradigm that allows both process and product. To interrelate the ideological perspectives of cognitive process and textual studies, Eiler could have pointed to ways in which the differing perspectives could lead to different conclusions unless they were considered simultaneously. Eiler makes some statements about a new instructional paradigm, but she could have pushed the point of needing a process and product alliance by citing colleagues associated with an individual perspective. Curiously, Eiler does not focus her discussion on similarities and differences between writing in the academy and writing in industry, although her findings are interesting. Her argument, that goal and genre matter to the act of composing, is substantiated in studies in the classroom and in the workplace. In a climate that often demonstrates differences between classroom and workplace writing, studies that point to similarities could be very important to framing the relationship of those two major divisions of writing practice.

Study 3: "A Case Study of One Adult Writing in Academic and Nonacademic Discourse Communities [Anna]"

In "A Case Study of One Adult Writing in Academic and Nonacademic Discourse Communities," Doheny-Farina (1989b) develops arguments for the validity of his study by linking the research to problems addressed by researchers in writing in the disciplines and in nonacademic writing and by focusing on a writer who is being initiated into two worlds of discourse. He assumes that both groups will be receptive to a case study of a writer learning to write in each setting. He can be confident that his ethnographic approach to the study of a writer writing will be acceptable methodologically to both groups as he does not defend his research methods.

But Doheny-Farina (1989b) does argue that academic and workplace writing can be researched simultaneously, using Perelman's concept of *institutional* discourse communities to link the two:

> Thus, while my analysis focuses on a unique individual writing in unique
> discourse communities, I will argue that in both communities the writer is
> producing institutionally based prose. That is, I will ultimately argue that
> it is neither the academic nor the nonacademic setting that distinguishes
> the types of discourse produced but rather the institutional role that the
> writer plays in each of those settings that fosters different types of dis-
> course. (p. 18)

Doheny-Farina obviously assumes that while researchers typically address either academic or nonacademic discourse, they can be convinced
to see similarities and insights gained from contrasting a writer's responses to the two settings. Thus he argues directly for the applicability
of both settings to both research groups.

In his initial efforts to establish validity, Doheny-Farina invokes
individual, textual, and the social perspectives but allies himself with
the social ideology. His authorities from the workplace intermingle
individual process research (Selzer, and Odell and Goswami) with textual studies (Freed and Broadhead) with social context studies (Odell,
and Paradis, Dobrin, and Miller); his authorities from academic writing
represent textual/genre perspectives (Comley and Scholes) and a blend
of genre and individual perspectives (Herrington). Thus Doheny-Farina
introduces all three of Faigley's ideological perspectives, assuming
that all ideological groups would find the topic he takes up worthy of
reading. In the analysis of the study, however, the researcher focuses
only on the social perspective through his discussion of normal and
abnormal discourse. He argues that newly initiated writers like Anna fit
into the institutional roles they are given, even if they resist them (as
Anna completed the clinic report but refused to put her name on it).
Anna writes normal discourse to maintain the community in the clinic
internship and abnormal discourse to change the community in the
literature class. Because many would claim that the nature of the workplace is to ask for writing that maintains the institution, and his study
of Anna stands as backing for that claim, Doheny-Farina asserts that all
institutions use both normal and abnormal discourse, backing his assertion with references to many other studies. He blurs the conclusion of
his case study of Anna, perhaps to combat a conclusion that potentially
would undermine the status of those who research professional writing.

Doheny-Farina's ideology of research is also shown by his treatment
of the individual case. He does not focus on the individual's composing processes, but rather on the individual's interactions with two

institutions and on the individual's differing writing roles in those two institutions. "The key difference between Anna's experience in both discourse communities," he writes, "lies not in the texts that she produced; nor does it lie in the communities themselves. The key difference lies in the relationship of writing to each community" (Doheny-Farina (1989b, p. 38). Had he been deploying the mixed ideology he referred to in the literature review, the researcher might have focused some of the analysis on texts produced in the differing situations, trying to sort through textual relationships, or he might have focused some of the analysis on the actual composing process sessions, trying to identify differences in authorial purpose or strategy that guided a composing act.

To reaffirm his social perspective during analysis, he turns to the social constructivist views of Bruffee and Rorty from Kuhn about the functions of normal and abnormal discourse in institutions. Their views are used to frame social/institutional implications both for Anna and for the gap between academic and nonacademic discourse. Within this theoretical framework he builds a social interpretation, focusing on the role-playing set by the institution and showing how both institutions contain both types of discourse but have different privileging systems. This institutional conclusion about the reciprocal relationships of types of writing reveals his allegiance with the discourse community of researchers and theorists involved with writing in the disciplines (Bazerman and Myers) and researchers advancing similar arguments in nonacademic settings (Paradis, Dobrin, and Miller; Doheny-Farina; and Harrison). This ideological comradeship also shows that he collapsed this group into one list, dissolving the academic and nonacademic dispute in the process.

But having invoked this social perspective, he does not become theoretically involved with the literature that describes how an individual is initiated into discourse communities either in nonacademic settings (there it focuses on internships, e.g., Anson and Forsberg, and learning organizational culture, e.g., Mathes) or in writing in the disciplines (there it focuses on learning genre or academy practices, e.g., Berkenkotter, Huckin, and Ackerman). Instead he links his findings to pedagogy, perhaps because a number of essays in the collection take a pedagogical turn, and admonishes teachers to make students aware that there are discourse communities. He connects the pedagogical turn to social ideology when he points to a call for " 'in-depth, comparative, ethnological stud[ies] of discursive practices in and out of school' " (Doheny-Farina, 1989b, p. 7) by Bizzell. This quote serves several purposes: It connects the academy and the workplace, it calls for the type of research reported in the study, and it calls for a social perspective on that research.

Doheny-Farina offers evidence for the internal validity/credibility and reliability/dependability of his study by describing the triangulation used in the two settings he contrasts in this case study—observation; field notes; discourse-based interviews; and open-ended interviews with Anna, her teacher, and her employers. He articulates those methods and then also builds more than one voice (see Clifford, 1983, on dialogic or polyphonic voice in ethnography) into his interpretation of the unfolding writing projects. In the discussion of the literature paper, he contrasts Anna's view with the teacher's or his own. He also adds independent evidence beyond his summary, detailing the teacher's plan for research papers, exchanges between Anna and the teacher, and quotes from interviews—all social constructions of reliability. The same use of multiple sources and voices is evident in the clinic case, in which he enriches his own interpretations with those of Anna and the supervisors in the clinic, quoting the discourse-based interviews when they articulate some of Anna's ethical dilemmas. His ability to reconcile a number of sources and voices into an interpretation of Anna's writing experiences makes the account more credible. Because his study employs a subject acting in two different environments, he could have made an overt case that the use of the same subject writing in the same time period enhanced the reliability of his comparison of the environments; that argument is available in the design. He could also have discussed why Anna was chosen as the subject as a way to assure readers that his subject was a reasonable choice. Both of these arguments would have enhanced his reliability arguments.

Conclusion

In each of the studies we have examined—Couture and Rymer (1985), Eiler (1989), and Doheny-Farina (1989b)—validity and reliability have been used as socially constructed persuasive arguments to help these researchers construct new understandings within their discourse communities. These arguments, especially from validity, also betray the studies' underlying ideologies. Couture and Rymer call on social ideology, but also use constructs from individual and textual perspectives. Eiler uses a predominantly textual ideology, but includes an experimental study focused

on individual composing choices. Doheny-Farina invokes multiple perspectives initially, then uses the study of one writer to articulate institutional questions about writing, from a social perspective.

Most researchers studying writing in the workplace are aware of
social forces that shape workplace texts and their studies of those texts.
Certainly the contributors to this volume are so aware. But little metadiscourse on research methods has raised questions such as the following: What means, if any, are preferred for establishing consensus among
this research community about what is quality research into writing in
the workplace; what roles should the traditions of consensus in empirical research (terms that include validity, reliability, credibility, dependability, and others) play in the construction of consensus in writing about
the workplace research? As we have demonstrated in this chapter, the
beginnings of these metadiscourse discussions are already lodged in
arguments used to establish validity and reliability in workplace research texts.

PART II

Pedagogy and Practice

10

Collaboration and Conversation in Learning Communities: The Discipline and the Classroom

JONE RYMER

Overview

Social constructionist theories have exerted a growing influence on our disciplinary conversation about teaching professional writing for well over a decade. We began by questioning positivism and individualism as our informing perspectives on professional writing. Influenced by Vygotsky, Kuhn, Toulmin, and others, we moved toward a social constructionist view of knowledge, defining discourse as a social phenomenon and according it a central role in *making meaning* in science, technology, the professions, and business. Over time, we rejected classical rhetoric's view of scientific language as a conduit for facts, and our instructional function as fostering

AUTHOR'S NOTE: I thank those who have engaged in the conversation about this paper: Charlotte Thralls, Nancy Blyler, Barbara Couture, John Beard, Janis Forman, and Douglas Kevorkian; and I thank Hugh Cannon for his support and Anne Broderick and Sylvia Ackroyd for their assistance.

the skills to convey those facts clearly and unobtrusively. We expanded the domain of professional writing classes to include the communal act of inventing *what* to say, as well as how to say it, of negotiating meaning among the members of the discourse community (C. R. Miller, 1985; LeFevre, 1987).

But applying a social constructionist view of knowledge to our pedagogy posed many problems. If we were to teach students the social practices of making meaning in disciplines foreign to us, disciplines in which writers typically conceived of their tasks in the positivistic tradition of "letting the facts speak for themselves," what was our appropriate role as instructors and what kinds of assignments were we to devise?

Many instructors responded by assigning *cases* and group writing tasks, activities reflecting a social perspective. Looking back, we now believe that while these tactical changes pushed our pedagogy well beyond positivistic and individualistic conventions, they are not fully compatible with the broader assumptions implied by a social-epistemic rhetoric: They constrain students' meaning-making as a social process, and they focus classroom discourse on the instructor teaching individual students the writing skills for professional employment.

Now, influenced by Rorty, Geertz, Bakhtin, Foucault, Freire, and others, our disciplinary conversation has taken another turn. We are transforming our pedagogy, overtly professing a social perspective that our knowledge of the world, ourselves, and others is constituted in written and oral conversation within communities. Assuming both discourse and education to be inherently collaborative meaning-making activities, we are now developing classroom strategies to reflect this new social rhetoric. Rather than limiting our mission to preparing students to learn the discourse practices of their chosen professions, we are openly presenting a social-epistemic perspective that language use is a dialogue for which we are all responsible. We are expanding our objectives to assist students' development so that they might fully participate in that dialogue—intellectually, socially, emotionally, and ethically. Our mission is to help them become not only effective and efficient practitioners but critical and responsible participants in their professional discourse communities (see Bizzell, 1986b; Cooper, 1989; Giroux, 1983).

Congruent with these expanded goals, we are transforming our classrooms into communities. Instead of attempting to transmit knowledge to students, we are inviting them to become co-learners with us, what Freire (1968/1983) called "teacher-student with students-teachers" (p. 67).

Thus we are relinquishing our traditional roles as authority figures, performers, and masters guiding apprentices, so that we can collaborate with our students, engaging them in conversation and acknowledging that our talk shapes our reality as a social group and, in turn, our pedagogy (G. Clark, 1990; Fox, 1990; Trimbur, 1989).

In this chapter I attempt to provide a perspective on this evolution in our teaching of professional writing. First I describe how our most common social practices have tended to countermand a social constructionist view of knowledge. Then I discuss how we are developing new strategies for learning that more fully realize a social constructionist rhetoric, strategies that are transforming our instructional roles so that we may become collaborators with students, consultants to them, and facilitators of their learning. Finally, in the spirit of contributing *local knowledge* to our disciplinary conversation (Geertz, 1983), I will describe my own classroom experiences as explorations in developing a learning community.

Assigning Cases and Writing Groups: Early Practices Reflecting a Social Perspective

During the 1980s, most professional writing instructors adopted some classroom practices manifesting a social perspective, especially cases for writing assignments and writing groups for involving students in reviewing texts. Although both methods emphasize a social component in the writing process—context in cases and discourse interaction in writing groups—neither represents writing as a fully social process of making meaning. Moreover, neither helps dethrone the authoritarian instructor in the learning process.

Case Assignments

Derived from the concept of the organizational discourse community (Faigley, 1985), workplace simulations in cases were adopted by instructors wishing to emphasize writing's social context. Cases gave students an opportunity to experience some of the workplace's complexity and its impact on composing (Couture & [Rymer] Goldstein, 1984, 1985). But there are limitations. Cases inhibit writers' full involvement in making meaning and thus cannot fully satisfy a social perspective on teaching writing. Even the best of cases lack reciprocity of writers and readers, and allow the writer no engagement in social processes, no influence on the context, little chance to invent meaning, and only a limited number of contextual constraints to guide composing or judge the product's effectiveness (see Mendelsohn, 1989).

Because cases are texts, readers must interpret them. Still, the mimetic power of cases can seduce students into considering them as microcosms of the real world, thereby encouraging instructors to function as interpreters for students who lack pertinent job experience, or worse, as classroom arbiters of the "correct" response to the case text (see Scharton, 1989). In fact, because cases are poised precariously between the academic and professional discourse communities and may seem to privilege students with job experience similar to the fiction, they may encourage instructors to assume authority over professional contexts. By imposing an academic interpretation, however, the instructor not only may discriminate against students with career experiences paralleling the case but may suggest an erroneous notion of universal criteria for effective writing in the workplace. In short, although cases offer many advantages as exercises in the professional writing classroom, their inherent limitations discourage a fully social pedagogy.

Writing Groups

Popularized by the collaborative movement in composition championed by Bruffee (1973, 1983, 1984), writing groups were the classroom method instructors adopted most frequently. Although based on the notion that writing is a social process, both peer response groups and collaborative writing assignments can contradict a social perspective by designating the ultimate responsibility for the construction of meaning to the individual writer and by locating authority for classroom discourse with the instructor.

Peer writing groups, offering audience feedback throughout the writing process, can help students focus on writing as a transaction with readers (Gere, 1987). Because of this social orientation and because peer groups have been found more effective than traditional methods for teaching individuals to write (DiPardo & Freedman, 1988; Gere & Abbott, 1985; Nystrand, 1986), many instructors have adapted peer response groups to prepare students for the workplace (Beard & Rymer, 1990; Louth & Scott, 1989).

Although students can learn about writing as a social process by discussing and responding to each others' texts, the dominant model of the peer review group is centered in writing as an individual act: Responsibility for creating meaning and decision making resides with the individual author; members have no stake in each others' processes or texts, so there is no exigency for negotiating communal understanding. Not only is the goal of peer response groups to improve an individual's skills but writers are not exposed to the diversity of viewpoints characteristic of the discourse community. Instructors—influenced by research supporting structured groups (Hillocks, 1986)—tend to guide students' responses by checklists, discouraging them from fully engaging the text themselves, substituting the "correct" perspective for their own reactions (DiPardo & Freedman, 1988; Nystrand, 1990).

Professional writing instructors' main motivation for adopting writing groups, however, has been to prepare students for the group authoring experiences that will characterize their career writing (Ede & Lunsford, 1990). Thus the trend has been away from peer review groups and toward group assignments calling for a "shared document" (Allen, Atkinson, Morgan, Moore, & Snow, 1987)—communal tasks that resemble those in the workplace (Lay & Karis, 1991; Louth & Scott, 1989; Tebeaux, 1991). Replicating workplace group-authored tasks in the classroom is unlikely to be effective social pedagogy, however, because although developing consensus can be a significant goal, productivity (rather than developing meaning or achieving broader educational objectives) tends to drive multiauthored arrangements in the classroom no less than the workplace (Van Pelt & Gillam, 1991). Such common configurations as division of the text into sections and allocation of topic components by expertise get the job done efficiently; they do not foster engagement among the participants in the collaborative construction of meaning or their development of the skills pertinent to the social processes of professional writing. In short, while group writing assignments can be designed to be conducive to a fully social pedagogy (see below), they can result in the instructor as manager of team projects in which each student functions mainly as an individual writer, attempting to demonstrate his or her effective skills.

Transforming Professional Writing
Classrooms Into Learning Communities

As we subscribe more fully to a social constructionist perspective and consider its wider cultural implications, we are embracing a broader mission and reshaping our pedagogical practices and our roles. *Collaborator* embraces the spirit of all our acts in transforming our classrooms into learning communities, but instructors should also play other roles, especially as consultants and facilitators. While these roles overlap occasionally, they represent distinctive functions manifesting a social constructionist perspective: *Collaborator* defines the reciprocal relationship necessary for engaging students in conversation, *consultant* defines a voice to speak as an expert on students' discourse practices, and *facilitator* defines a context in which to act, constructing favorable circumstances for students' own learning.

Transforming our classrooms into learning communities fully compatible with a social-epistemic rhetoric is a task that each of us must ultimately pursue with our students, developing a pedagogy pertinent to our own contexts (Chase, 1988). Thus in discussing potential roles for instructors of professional writing, I will suggest ways we might develop our practices by using my experiences in the business school at a large urban university (Wayne State in Detroit), where the typical undergraduate is an older (age 28), long-distance commuter, who works many hours a week, often in a full-time career position. This discussion is in three parts: first, how we may collaborate with students in the learning community; second, how we may act as consultants to them; and third, how we may facilitate their learning.

Collaborating in a Learning Community

As collaborators in a learning community, we can participate in classroom conversation, interacting with our students and limiting our tendency to give answers and render judgments. This role does not mean that we abdicate our responsibilities as instructors or assume the relativist posture that one view is as good as another. Rather, we accept learning as a dialogue, a mutual and complementary creation of knowledge (G. Clark, 1990), and we accept "as legitimate the authority created by collective discursive exchange and its truths as provisionally binding" (Bizzell, 1990, p. 665). When we adopt this collaborative role, we teach our students by *how* we teach, modeling participation in a nonauthoritarian community by enabling genuine dialogue, sharing authority, and mutually inventing a fuller classroom discourse (Cooper, 1989; Hynds & Rubin, 1990; Peterson, 1991).

Enabling Genuine Dialogue

As co-participants in the classroom, we can advocate genuine dialogue, but how can we enable it? Establishing a community that engages in authentic interaction first demands students' active participation as whole persons. In assigning peer reviews, for example, we can invite students to respond as readers with their own ideas and feelings, and then to develop interpretations of how target audiences might read the text in context. Rather than dictating their opinions through checklists that accomplish our objectives (Nystrand, 1986, 1990), we can demonstrate our commitment to their genuine participation by simply respecting their own responses as the basis for their developing an understanding of how readers read.

To extend students' conversation among themselves to engage in dialogue with us, their instructors, we can model for them. By our actions —listening as well as telling, learning as well as teaching, becoming persuaded as well as persuading—we can encourage them to tell, to teach, and to persuade, and thereby engage in conversation that is truly reciprocal. To converse authentically, however, we must *need* to listen and to learn from them. In classrooms where many students are career employees, for example, students can create case assignments based on their own workplace writing. Assisting in the production of such cases, the instructor becomes a collaborator with the case developers and a fellow learner with other team members, exploring writing processes and contexts in new organizational cultures. In contrast to published cases, the student-generated case represents a collaborative venture in which expertise and instructional roles are shared by students and the instructor, thereby modeling the learning community. Moreover, student-generated cases can provide access to fully contextualized, dynamic workplace situations for assignments, permitting group members to engage in writing as a social process of making meaning, thereby attenuating drawbacks of published cases such as the absence of reciprocity between writers and readers.

Sharing Authority

Authentic dialogue among students and instructor on such a common task can begin the transformation of the classroom into a learning community. But ultimately, if we are to encourage students to drop the mask of "student" and speak genuinely as full persons with social histories (Bleich, 1988; Fox, 1990), the instructor may need to relinquish traditional professorial authority in a dramatic fashion. By openly sharing responsibility with students for forming groups on the basis of diversity, the instructor can involve the whole class in a critical decision students deem of such importance that even the most reluctant participate in the negotiations.

A joint process on a sensitive issue like diversity of group membership represents risks no matter how narrowly students might define their potential differences. (Among my students, gender, race, age, career and family issues are significant while social class, religion, and most ethnic identities are generally ignored.) Engaging students in the formulation of collaborative groups not only demonstrates the instructor's commitment to sharing authority but both models and opens opportunities for genuine dialogue about a significant issue influencing the consensus-building process. Students must consider the identities of group members as part of their agendas, so that they can begin to explore these differences rather than avoid them as irrelevant to their conversation and problem solving. For example, when a class began discussing the value of diverse perspectives in task groups, one of my students said she felt as if she were going to be "bused across the room to a white group." Later, she admitted: "I was afraid to ask 'other' students, so I would've just formed a group with [three black women]. However, I know when I go to work, it won't be just with 'black ladies.' "

Inventing a Fuller Classroom Discourse

Once students have openly confronted their various social identities, they can collaborate with the instructor to articulate their goals on this issue: to develop an awareness of how they can participate in a professional or business community and how their individual skills and identities—their social histories, personalities, feelings, and values—might influence their contribution to that conversation. Then the instructor can facilitate their examination of the interplay among their lives and their educational experiences, helping them forge a mode of discourse that can deal with differences directly and respectfully. Formerly, for example, some of my students were plainly embarrassed by the highly rhetorical style and bold gestures of those who had learned public speaking in black evangelical churches. They could not value the skills of this "alien" discourse community or see the relevance of these skills to the classroom and professional contexts. After confronting differences in classroom discourse, most students reacted to this phenomenon supportively with ways their peers might adapt their abilities to the business world.

Such experiences help participants invent language to bridge differences, constituting themselves as a group shaped by and benefiting from multiple viewpoints. In turn, the conversation helps each member shape his or her professional identity by critically negotiating among the social definitions offered by different communities, as well as the individual's self-image (Brooke, 1991). Rather than viewing the adoption of an appropriate persona as leaving one discourse community for another (shedding the colorful rhetoric of the evangelical church for the bureaucratic style of business, for example), students can be encouraged to see themselves as members of multiple communities. Accordingly, they must each choose for themselves how fully they will embrace the classroom community and, in turn, the language, conventions, values, and assumptions of the professional discourse communities they will join (Bizzell, 1986b; J. Harris, 1989).

Consulting With Members of a Learning Community

In contrast to the role of collaborator, *consultant* defines an instructor's role as communications adviser on students' discourse. Acknowledging our status as experts and critics, we can advise, challenge, and attempt to persuade students to develop wider perspectives, but we should respect their authority over their own communications. After discussing consulting roles as expert and critic, I will describe how, acting as a critic, I have attempted to deal with one of the more sensitive problems frequently occurring in my students' groups.

Consulting as Experts

The role of consultant provides us with a way to bridge academic and professional discourse, freely admitting that we are not members of our students' work groups or their professional communities. Acting as consultants, we can help students develop the capabilities to master the practices and "read" the contexts of professional discourse communities. As consultants we can advise groups, enabling them to manage their own enterprises—an essential condition for learning about writing collaboratively. Barriers to effective writing groups—extremely complex and only partially researched—can stem from typical group problems like poor conflict management, as well as from the composing process overlaid on the interaction processes (Forman, 1990; Forman & Katsky, 1986). Instructors can avoid managing students' groups on the one hand, and fostering mere trial-and-error learning on the other, by boldly acting as consultants who represent a safe source of informed advice and counsel about composing and group processes.

Consulting as Critics

The consultant as expert is complemented by a more radical role, consultant as critic. Because we are not members of our students' groups or professional discourse communities, we can speak as the outsider— for example, challenging a group's process when the members meekly confirm the status quo instead of acknowledging a risky consensus. In effect, the role of critic obligates us to provide the outsider's often disturbing perspective, even calling into question those basic assumptions that our students, as potential or actual stakeholders, tend to assimilate along with their initiation into the discourse conventions.

Instructors who act as critics in this broader definition of consultant —including counseling both groups and individuals, and, as a last resort, directly intervening as facilitators, not managers—can help students use language not only to reproduce the community but also to critically examine it, thereby gaining some understanding of how their conversation creates the meaning in their communications and the very reality of their groups (Putnam, 1986; Rafoth, 1990). For example, student writing groups do not necessarily operate in a truly collaborative fashion, democratically achieving consensus through fully reciprocal dialogue that reflects the views of all members. In some groups, consensus is displaced by the hierarchy of power to which members are accustomed in their jobs; in many groups, a *false consensus* represents domination by the academic elite or a majority, not the mutuality and egalitarianism implied by collaboration. (The term *collaborative writing* has caused considerable debate and misunderstanding, see Forman, 1991, 1992.) The privileging of consensus thus can suppress individual views (as many have charged—for example, Bizzell, 1986b; Fox, 1990; Myers, 1986), especially marginalized voices—in my experience at Wayne State, minorities, immigrants, and most of all, the working class. How can instructors deal with divisive issues like social class that prevent some students from participating fully in shaping the consensus?

Dealing With Sensitive Issues:
Social Class

Although we are inventing classroom discourse to initiate dialogue on racial, multicultural, and gender issues (for example, Lay, 1989), we do not have language to speak about such sensitive issues as social class. Because none of us wants to admit that class divides us as a people—most of us consider ourselves middle class, simultaneously sharing the values and norms confirmed by the ethos of business and education, while subscribing to the myth of *classlessness* (DeMott, 1990)—we as instructors may be unaware that diverse groups with a minority of working-class students frequently adhere to a hierarchy governed by middle-class values, norms, and interpretations. We may be unaware of possible class biases in our assignments that support that middle-class hegemony. For example, extended group projects burden working-class students whose jobs demand their primary allegiance, and whose values and study habits (for example, *satisficing* instead of doing their best) may disadvantage them compared with middle-class members whose priority is getting good grades. We may be unaware that neither we nor our students have ways to talk across the barrier separating angry middle-class students from the single parent who works the night shift: In the middle-class members' eyes, she is irresponsible for staying home with her kids instead of attending a meeting in one of their suburban homes; in the working-class student's eyes, it's her one night off, and "they" never listen to her anyway ("They keep voting me down"). These are peers who live in different worlds, although typically neither they nor their instructors can even acknowledge the fact.

The role of the consultant-critic in dealing with class conflicts is fraught with risks, but we should not avoid class issues by narrowly defining our roles as communication specialists in the tradition of positivism. Class conflict, rather than encouraging a broader consensus and developing members' awareness of diversity, can increase the alienation of all group members. Middle-class students who eventually stop listening to a working-class member in their midst often do so in the name of democratic precepts ("Majority rules," they say). Violations of middle-class workplace norms ("Be on time" and "Call if you can't come") convince them that they are justified in ignoring the irresponsible member's views, isolating her from the consensus-building process.

Acting as consultant-critics, we have the *possibility* of guiding our students to become critical, aware participants in their discourse communities—voices capable not only of speaking effectively and efficiently but of speaking with respect for others while also speaking out, challenging assumptions and forging a broader consensus. But sometimes, despite our best efforts, the dialogue fails and the best we can do is keep the conversation going, telling the stories of our lives. At those times, using a *rhetoric of dissensus,* we as consultant-critics can help students seek consensus "not as an agreement that reconciles differences through an ideal conversation but rather as the desire of humans to live and work together with differences" (Trimbur, 1989, p. 615). In short, we should strive to develop a sense of community that does not depend on conformity—instead, as Rorty (1989) says, we should try "to extend our sense of 'we' to people whom we have previously thought of as 'they' " (p. 192).

Facilitating Students' Learning

As facilitators, we can develop activities to enable students' active conversation among themselves, thereby fostering the circumstances for learning collaboratively across a wide range of objectives. Some guidelines for constructing collaborative contexts for learning seem self-evident: For example, does the task empower students to conduct the activity themselves? Other essential criteria for framing students' conversations may be less apparent: Do assessment systems involve the students and focus on social practices defined by our objectives? Do assignments foster these practices, especially the engaged conversation that creates meaning and supports the community?

Of paramount importance to achieving the goals of a social pedagogy is grounding all practices in the mutual responsibility of the individual student and the group/community. After briefly considering the special problem of assessment I will discuss how assignments might foster making meaning through individual and group authorship, and then conclude by describing a videotaped task in which my MBA students and I have applied these ideas in the classroom.

Constructing Assessment Systems

Typically positivistic and individualistic, assessment would seem to represent the antithesis of a social constructionist pedagogy. Grading designates us as authorities who must measure the amount of knowledge each student has acquired (Bleich, 1988). Although mandated grading presents an impediment to collaborating with our students, it is not an insurmountable obstacle—if we conduct it as a collaborative, meaning-making activity.

Instructors have suggested that group writing should result in both individual and team grades (Bosley, 1990; Morgan, Allen, Moore, Atkinson, & Snow, 1987), but giving each student multiple grades represents only the first step in matching our social mission with appropriate assessment. Not only should we enable students to participate in the assessment process but we should facilitate as well as evaluate their development on a wide range of nontraditional criteria, such as demonstrating interpersonal sensitivity, facilitating group interaction, and supporting others' learning. How can we facilitate mutual teaching and learning, for example, while fairly rewarding each student?

By constructing an assessment system that holds both the individual and the group collectively responsible for each student's development, the instructor can empower students to teach and learn from each other (Slavin, 1990). Like all aspects of our learning communities, assessment should be the subject of a dialogic process involving the students (through journals, for example, [Rymer] Goldstein & Malone, 1984, 1985), and it should be adapted appropriately for our complex social goals, not merely isolate a section of a product to represent an individual's writing. Appropriate evaluation for a social pedagogy can approximate a miniethnographic study: observing, collecting evidence, using various methods to record participants' perspectives in depth, eventually building a *thick description* of a group's experience that can represent some of the complexity of professional writing as a social process. Although such assessment is lengthy and cumbersome, sensitive and partially secretive (one must not breach the confidentiality of any group member), it can bring students into the process and allow instructors to facilitate and to fairly reward both the group and each member's participation on a wide variety of criteria (Beard, Rymer, & Williams, 1989).

Developing Assignments

As facilitators, instructors can help students participate in an engaged conversation wherein dialectic is integral to consensus building. The key to this engagement, however, lies in designing assignments that can facilitate the substantive conflict requisite for making meaning. If participants create a *true consensus* through the interaction of their varied perspectives, they might experience the dialogic character of the discourse community. A *false consensus* can be based not only on a majority silencing a minority (like my working-class students) but also on privileging compromise and agreement.

Because collaboration suggests compromise as favored by Rogerian rhetoric's focus on common interests and shared understanding, our collaborative assignments can inadvertently support group members' misconception that achieving consensus does not entail airing intellectual differences (Karis, 1989; Myers, 1986). In fact, by stressing collaboration in our assignments, we can unintentionally encourage group members' tendency to avoid conflict at any costs (Wall & Nolan, 1987).

Both theory and research suggest, however, that dialectic is essential to the conversation that would create meaningful consensus. Rhetoricians note that dissonance is a necessary condition for discourse (Bitzer, 1968). Communication scholars warn of the dangers of *groupthink,* insisting that conflict is requisite for the best solution (Janis, 1972), advising participants to differentiate interpersonal from substantive issues so that they won't shun the latter (Putnam, 1986). Professional writing scholars now advocate intellectual conflict, encouraging consensus through the full play of ideas in dialectic, what Burke called the "competition which leads to cooperation" (quoted in Karis, 1989, p. 120). Research on professional writing tentatively supports this view: Less effective documents may result when groups fail to engage in substantive discussion of rhetorical issues and document content (Burnett, this volume; Rogers & Horton, 1992).

If dialectic is a condition for members of a discourse community to achieve insight and create meaning (Morgan & Murray, 1991), our assignments should facilitate the competition of ideas. But how can we encourage the necessary commitment and intensity that will enable all members to see intellectual negotiation as part of building consensus? One answer is to construct writing assignments derived from Bales and Strodtbeck's (1951) classic criteria: interdependent work in which all members have a significant stake in the outcome. Applied to writing tasks, these conditions imply a synthesis of communal and individual responsibility in authorship (Farkas, 1991; Morgan, this volume). Although a *shared document* lends itself to group ownership, individual stakeholder status for each member is not inherent in the multiauthored task (as discussed above). Assignments must be structured to depend on individuals assuming mutual responsibility with the group so that all members will engage in the debate of ideas and construct a true consensus. An assignment built on this communal and individual interdependency and authorial responsibility is illustrated in the following videotaped task.

Empowering Students Through a Videotaped Task

On the basis of theory and classroom experience, instructors can devise methods empowering students to conduct their own tasks, to participate as individual and collective stakeholders in authorship, and to share mutual responsibility for facilitating and assessing their own growth. For example, videotaping a group writing assignment that fosters engaged debate can empower students through the images of their authorial interaction to reflect on their processes and evaluate their participation, thereby facilitating the group and the development of each member.

The task for the videotaped session asks team members to plan together a short document in response to a controversial case. After the group debates strategies and develops a consensual solution, each member writes alone—as if this draft were the sole communication. The group then reviews the drafts, selecting those that they could approve. Videotaping is integrated into the assignment by taping the group planning session but interrupting it before the members achieve consensus. Immediately viewing the "interrupted" videotape enables the students to revisit their planning interaction, analyzing and adjusting their solutions and strategies while they are motivated to air their views fully and thus draft a document reflecting the team's consensus.[1]

Adapted from the workplace approval system in which writers are accountable for representing the views of others as well as their own (Couture & Rymer, 1989; Paradis, Dobrin, & Miller, 1985), this assignment makes each participant a stakeholder, encouraging sub- stantive interaction on all issues of composing and, therefore, a genuine consensus that each can articulate and support. The task clearly defines the "right" response of each writer as representing the group's consensus, the communal meaning each has helped create and can evaluate. In addition, the video and drafts, together with written critiques of the conversation, make each student accountable to the team and, in turn, the team accountable for the learning of all the members.

Reviewing the videotapes of their planning while in process enables team members to assess their own and their peers' skills as an integral part of facilitating their work. The videotapes are images of their interaction that, when reviewed under the task's duress, facilitate members' perception of unresolved issues, lost perspectives, and the failures of negotiation—in short, how their conversation constitutes the consensus and the reality of their group. These visions then become revisions when (with the instructor acting as consultant) members individually review the tapes later to assess their own participation.

For some teams, the videotapes clarify the difficulties in developing and articulating the team's consensus, thereby enabling students to understand the essentially dialogic character of discourse as it applies to such practical matters in the workplace as assigning writing tasks (Couture & Rymer, 1991). For others, the videos can help them confront the hegemony from which their consensus derives, developing awareness of the majority's tendency to build consensus by excluding minority voices. In the videotaping of an MBA group with an international member, for example, the students engaged in some intense discussion that represented a collision of values in analyzing the rhetorical situation: The native students were primarily concerned about their company's allegiance to democratic values (freedom of the press), whereas the nonnative student empathized with the audience's sense of public embarrassment. Eventually the nonnative agreed with the majority view so that the members avoided a full airing of the issue and expressed relief at achieving "consensus" ("We all felt the importance of making [K——] feel comfortable with the consensus opinion"). In reviewing the video, however, one member questioned the majority's easy assumption that democratic ideals should obliterate any concern for individual feelings:

> I thought it was particularly interesting that K——, who was brought up in a different culture, was the only person who seemed sympathetic to the customer and questioned whether we were in the right. . . . I wondered if the American culture was the cause of the remainder of us being more hardened or cruel about what should be published.

For this group member, the video image of her group so confidently yet so "sensitively" moving against the minority view provided a personal epiphany. When incorporated into her written critique, this epiphany became the vehicle for all team members to reflect on their actions. In sum, while these MBAs were learning to assume the conventional ethos in American business, they were learning and teaching from each other to critique "democratic" procedures and question both group and individual assumptions about communal meaning and the consensus-making process.

This assignment facilitates dialectic and individual/communal responsibility through the structural design of the task combined with technology. Intensifying members' involvement while distancing them, the tapes offer students a vision of their social construction of meaning and foster the development of critical perspectives on their discourse practices and assumptions. Collaboratively viewing and reviewing the tapes—records of students' growing potential for creating a richer, more inclusive consensus—affords instructors opportunities to consult with students and to learn from them.

Thus do we edge toward becoming "teacher-student with students-teachers," members of an authentic learning community. But the goal of creating a true classroom community is often elusive, and frequently our circumstances cast us in the old roles of knowledge dispenser, manager, judge, and certifier, and sometimes we elect the familiar scripts ourselves. Building a community with students as fellow learners—participating with them in conversation, consulting with them, facilitating their learning—is a dynamic and delicate collaboration that we must continually begin anew.

Note

1. Videotaping groups in process risks replacing their genuine discussion with performance for the camera. If students are accustomed to videotaping (and if the interaction itself is not graded), the time constraint combined with the group/individual nature of this assignment will encourage their involvement with the problem solving and attenuate the impulse to perform.

11

Postmodern Practice: Perspectives and Prospects

RICHARD C. FREED

> The most transparent and hence invisible function of practice is the manufacture of belief—that is to say, the telling of interesting stories.
>
> —Richard C. Freed et al.,
> "Postmodern Practice:
> Perspectives and Prospects"

I should confess. Soon after accepting the invitation to write this overview of practice and professional communication, I realized that I didn't know, really, what practice was. I sensed that it was always process, always means *toward* some end but without an End, for the End of practice —or so I thought at the time—at once its desire and demise, is perfection. If practice makes perfect, that is, it ceases to be. Waxing philosophical, I therefore turned to the philosophers, to Aristotle and Bernstein and Habermas, who wrote about practice or *praxis* or both. This was a dead end.

It got worse. I didn't know who my audience was, or why they would be reading. More accurately, the potential readers were so varied in their interests and knowledge that I couldn't decide on a target audience to whom most of the discourse could be directed. Practically oriented practitioners? Research-oriented scholars? In-the-process-of-being-oriented graduate students?

But the readership problem, I soon realized, was only the symptom of a more fundamental one: I didn't even know what professional communication was. I knew the term, of course. I had read it in the journals, and I had spent the previous two years directly involved in writing a proposal for a doctoral program in rhetoric and professional communication. But somewhat like gravity, I confess gravely, though *professional communication* was all around me and I could feel it at every step, its meaning wasn't clear. There seemed to be some relationship between the term and other more established ones like *business communication* and *technical communication,* the latter two probably, at least in part, subsets of the former. The latters' longer tenure as terms, however, didn't clarify matters much, because one can compose technical communication in the business profession (those who do are often called *practitioners*) and business communication in a technical profession (those who do are not). More confusing, the related term *nonacademic writing* refers to discourse (e.g., proposals and reports) quite often composed by college administrators as well as professors in the academy.

Focusing on genre helped a little, because few people would consider novels one of the standard forms of professional communication but most would probably include reports and proposals. Genre, however, only went so far, as few people, I believe, would consider as professional communication a proposal written by a student to fulfill a course requirement. Even if the student, as part of a class assignment, wrote a "real" proposal in response to a real organization's request for proposals, the student would have practiced professional communication but not, it seemed, produced any. Only if the student submitted the proposal to the real organization would she have engaged in the practice of professional communication. I attempted the following provisional definition to help explain why.

Professional communication is

A. discourse directed *to* a group, or to an individual operating as a member of the group, with the intent of affecting the group's function and/or
B. discourse directed *from* a group, or from an individual operating as a member of the group, with the intent of affecting the group's function, where *group* means an entity intentionally organized and/or run by its members to perform a certain function.

Primarily excluded by this definition of *group* would be families (who would qualify only if, for example, their group affiliation were a family business), school classes (which would qualify only if, for example, they had organized themselves to perform a function outside the classroom— e.g., to complain about or praise a teacher to a school administrator), and unorganized aggregates (i.e., masses of people).

Primarily excluded by the definition of *professional communication* would be diary entries (discourse directed toward the writer), personal correspondence (discourse directed to one or more readers apart from their group affiliations), reportage as well as belletristic discourse— novels, poems, occasional essays (discourse usually written by individuals and directed to multiple readers not organized as a group), most intraclassroom communications (e.g., classroom discourse composed by students for teachers) and some technical communication (e.g., instructions—for changing a tire, assembling a product, and the like; again, discourse directed to readers or listeners apart from their group affiliations).

The definition didn't eliminate all the gray areas, especially in promotional discourse, but it did underscore this point: Professional communication is decidedly functional and eminently and complexly social. It would seem different from discourse involving a single individual apart from a group affiliation communicating with another such person, or a single individual communicating with a large unorganized aggregate of individuals, as suggested by the term *mass communication.* If the discourse is professional communication, then receiver or sender or both must be a group or an individual situated within the group. Given the number and diversity of such groups, the variety of their functions, and the complex of messages directed to or from them, professional communication is a very large discipline indeed.

What was needed, it seemed to me, was some discussion (part I, below) of the forces that inscribed the field on which current practice is played, some discussion of the history, the natural history as it were, that shaped the field, carved its grounds. Although only in the last decade has it begun to be mapped and charted, much of the terrain will be familiar to many scholars in rhetoric and in composition studies who understand that the grounds have not only shifted but are disappearing before our eyes. That terrain is being marked by a movement from determinacy to indeterminacy, from the totalizing to the heterogeneous, from the global to the local, and from knowledge as essence and other to knowledge as constructed and believed. Part II views this movement from the point of view of audience, arguing that audiences are composed of group-constructed roles that directly reflect the organization's or group's manner of constituting, storing, and transmitting knowledge. Part III examines the organization of organizations and how they organize knowledge, arguing that the global economy is now, and the corporate structures of the future will be, characterized by a decentered space of flows isomorphic with hypermedia data systems. Part IV comments briefly about the effects of all this on practice, in a world without end or beginning.

I

We need to start at the beginning (at least some beginning) when there was an end, beyond embodiment and toward which the human drama was enacted. This master narrative, as Jameson and Lyotard might call it, justified conduct and conditioned knowledge and belief. In making real the infernal, the supernal, and divine intervention, its hero, Everyman, the community of believers, struggled to stay the virtuous path to the city on the hill, despite the antagonistic forces of darkness, whether dragon or devil or forest of errour. After science eroded the power of this narrative, relegating it to the category of story and myth, two other master narratives, according to Lyotard (1984), rose in the 19th century to legitimate knowledge: the more political narrative of emancipation and the more philosophical narrative of speculation.

In the narrative of emancipation, humanity plays the role of the hero of liberty, the state receiving "its legitimacy not from itself but from the people. . . . The nation as a whole was supposed to win its freedom through the spread of new domains of knowledge to the population" (Lyotard, 1984, pp. 31-32). The dissemination occurred through state agencies (the higher educational system) and the professions (including scientific institutions). Access to science was a right of the people, no longer denied scientific knowledge by a state that functioned as or was significantly allied with a tyrannical priestcraft. In this narrative of freedom, the epic of humanity is the story of its emancipation, its struggle to defeat all that would prevent its self-governance (p. 35).

In the narrative of speculation, the subject is not practical and political—humanity as the hero of liberty, but cognitive and philosophical—the speculative spirit as the hero of knowledge. Higher education justifies its existence not by the criterion of usefulness, preparing the citizenry to perform its functions, but by unifying the now scattered and separate sciences (Lyotard, 1984, p. 33). Rather than science serving the interests of the state, knowledge "is entitled to say what the State and what society are. . . . The humanist principle that humanity rises up in dignity and freedom through knowledge is left by the wayside" (p. 34).

In both these narratives, science plays an important role, even though the language games related to scientific and narrative knowledge are incommensurable. Science's denotative game of distinguishing between the true and the false cannot validate the knowledge of narrative's prescriptive game (particularly evident in the emancipationist narrative), which distinguishes the just from the unjust. Because the criteria relevant to each game are different, science must conclude that narrative statements cannot be argued or proved. They are relegated to the "savage, primitive, underdeveloped, backward, alienated, composed of opinions, customs, authority, prejudice, ignorance, ideology. Narratives are fables, myths, legends" (Lyotard, 1984, p. 27). Despite these two distinctly different kinds of knowledge, however, science had to rely on narrative for its own legitimation because science could not "prove" itself to other than scientists, could not claim its knowledge as true, without recourse to epic. In the popular media, for example, scientific discoveries are always recountings, narrations, epical and sometimes epochal quests for the cure for polio or HIV, or the search for quasars and pulsars. Thus did science seek its legitimation through narrative, even while it devalued this form of knowledge so different from itself.

Because, however, these two narratives existed as alternative versions, the meaning of legitimation varied. A single narrative and, by implication, narrative itself was not capable of describing or validating that meaning (Lyotard, 1984, p. 31). For Lyotard, the delegitimation and consequent decomposition of the master narratives is very much a good thing.[1] The master narrative is representative of a totalized philosophical system that, like any narrative, constrains diversity and encourages homogeneity. Whether composed by an individual, an organization, or a culture, narratives are by necessity smoothed whenever rhetorical choices are made regarding who can speak and be spoken to and what is and can be said and done. From a political perspective, this authorial control is significant. Master narratives will always be repressive of some people or groups who are neither admitted to the plot nor capable of changing it. Those whose smaller narratives attempt to be emplotted in the whole, thus altering its perceived unity and totality, are considered subversive.

What happened when the universe, that unity and totality, became polyverse, when sameness became difference, when, that is, unitary conceptions of knowledge legitimated by master narratives became riven and splintered? The realization, according to Hayles (1990), "that what has always been thought of as the essential, unvarying components of human experience are not natural facts of life but social constructions" (p. 265). This realization occurred in the 20th century in a process Hayles calls *denaturing,* as the foundations supporting conceptions of language, context, and time all crumbled and then disappeared.

Language became denatured when it was no longer seen as a representation of objects in the world. Early in the century attempts were made in mathematics, physics, and language theory to ground representation in a neutral metadiscourse, a formal system unpolluted by subjectivity and therefore ambiguity. In mathematics the notion of a formal system, introduced by Frege in 1879, was considerably advanced by Whitehead and Russell in their *Principia Mathematica* (published in 1911-1913) and widely accepted in the 1920s. In physics, according to Hayles (1990), Einstein's special theory of relativity was an attempt "to provide an overarching framework within which observations from different inertial systems could be reconciled" (p. 267). In language theory, the logical positivists attempted to create an unsituated metadiscourse, an unmediated channel between the writer or speaker and the written or spoken about.

By 1931, however, Gödel had demolished the attempts at formalization, demonstrating that any formal system adequate for number theory (e.g., arithmetic) contains an unprovable formula and that the consistency of such a formal system cannot be proved within the system. Following World War II, Kuhn (1962) and other philosophers of science argued that the existence of a unified field presupposed no vantage point outside the field: The observer is never apart from what is observed and hence observational statements are never neutral but always theory laden. Similarly, de Saussure (1966) argued that words do not derive their meaning from objects outside the field of language that they might appear to refer to or represent but from the difference between them and other signs that comprise the field itself. In Derrida's (1976) famous phrase, nothing exists *hors de texte.* Texts do not refer to some neutral, stable, grounded, natural world of objects apart from the field of language, but obtain their meaning from their relational differences, their intertextuality, with other texts.

Context became denatured in large part through the work of Shannon at Bell Labs. Shannon's 1948 papers were instrumental in creating information theory, which relies on divorcing the informational content of a message from its context or meaning. If information were not so divorced, then its value would be subject to change every time that the context changed, just as our sense of the term *recess* changes across various contexts, for example those suggested by elementary schools, legislative sessions, and topography. Separated from context, however, information could be quantified and hence processed and stored efficiently and rapidly. As a result, information technology developed quickly.

If information could be separated from contexts, then informational texts could be separable as well, especially if *texts* are viewed in their larger sense, not necessarily as things written down but as action inscribed. What then become readable are not only words but also everything from social custom to genetic code. In the scientific domain, for example, genetic engineering often involves taking acontextual information from its "natural" site or context and splicing it at another. In the political domain, current notions of context control and the related issues of spin and disinformation are possible only if context is no longer universal, stable, and given, but socially constructed; that is, if information is no longer conceived (in the originary sense of the word) from preexisting contexts but exists rather as ungrounded input from which those contexts can be created.

Time became denatured with the dissolution of the metanarratives and their tightly structured plotting of human experience. Only a salvationist metanarrative could authorize a work like *Pilgrim's Progress;* only an emancipationist metanarrative, those *Bildung* common to the 19th century. In each, progress and development are teleological, smooth, and causal. Neither instance of legitimation would be likely to accept an argument like Kuhn's (1962) against the cumulative development of science or a theory like Gould's, on punctuated equilibrium, about the progression of evolution. Interestingly, Kuhn sees his notion of progress in scientific revolutions as analogous to that in evolutionary theory. Most troubling about the *Origin of Species,* to those who were troubled by it, was not the idea of evolution itself, which had been in the air for decades, but the unteleological character of Darwin's version, which recognized progress and development, but toward no goal (Kuhn, 1962, p. 171). Progress away from, not progress toward. Fragmentation, discontinuousness, undecidability, and indeterminacy—these are all functions of a reconceptualization of time as no longer plotted, no longer fixed (in story and in ground) to that which gave it a beginning and an end.

The denaturing of language, context, and time thus resulted from failures earlier in the 20th century to eliminate through unified theories ambiguity and self-reference. Although these attempts were not successful, they problematized the notion of representation and resulted (though cause and effect here are difficult to determine) in a condition of fragmentation, indeterminacy, and heterogeneity. The breakdown of the master narratives produced an explosion, like the big bang of creation, distributing *petit recits* throughout the industrialized world; marking the extraordinary multiplicity of contemporary consciousness, "the radical variousness," as Geertz (1983) writes, "of the way we think now" (p. 161); and generating a profusion of perspectives and a ferment of critical discourses, articulated by a wide variety of disciplinary matrices to use Kuhn's (1962) term or interpretive communities to use Fish's (1980). In this new situation, according to Geertz (1983), "at once fluid, plural, uncentered, and ineradicably untidy" (p. 21), " 'a general' theory of just about anything social sound[s] increasingly hollow," and "a unified science . . . has never seemed further away, harder to imagine, or less certainly desirable than it does right now" (p. 4).

As a result, those who have written about the postmodern condition tend to privilege, not the general and global, but the local dimension. We have seen this privileging clearly in Lyotard (1984), for whom the breakdown of the metanarratives and the totalized system they suggest precipitates the proliferation of little narratives. It also exists in Rorty (1979), for whom the monolithic theory of knowledge, founded on optical metaphors and an unfounded objectivity, needs to be replaced by a dialogue of local interpretations. It especially exists in Geertz, for whom all knowledge is local knowledge, shaped by the specific culture that generates and believes it.

That local knowledge exists as a representation, but only as a representation, rather than as a mirror of something given "out there." Such is not to say that the world we move around in and bump into is immaterial. Gravity, as Hayles (1990) argues, does indeed exist, but its *representation* differs, depending on the community of believers for whom it is a representation: for Newtonians, the result of the mutual attraction of masses; for Einsteinians, the result of space's curvature; for some native Americans, the result of Mother Earth's calling out to the kindred spirit of the falling body (p. 221). Even something apparently as foundational as common sense, Geertz (1983) demonstrates, varies from culture to culture. What is common is not universal; it exists, rather, because a community grew it, and its shape, temper, and tonality will be different elsewhere.

Because the shape and tonality of knowledge vary by locale, and because for that locale the tone and temper of its knowledge rings and feels true, truths at one locale may be different from those held self-evident at other sites and from those held at different times in the same locale. Therefore, the recognition of and the consequent emphasis on diversity and pluralism. This recognition is fundamental to Rorty's (1979) philosophy of conversation rather than confrontation. It is very much evident in Geertz (1983), whose enterprise seeks to understand "how the deeply different can be deeply known without becoming any less different" (p. 48). "We are all natives now," Geertz writes, "and everybody else not immediately one of us is an exotic. What looked once to be a matter of finding out whether savages could distinguish fact from fancy now looks to be a matter of finding out how others, across the sea or down the corridor, organize their significative world" (p. 151).

We will continue to examine indeterminacy, fragmentation, representation, the focus on the local, and other postmodernist issues as we consider recent changes in the organization and process of production, both globally and locally; in the view of organizations themselves; in the raw materials for discourse; in the idea of authorship; and in the idea of audience, to which we now turn.

II

As so often, locution is the sediment of belief:

"Have you thought about *your* audience?" said the editor to the writer.
"Yes. I've thought about *my* audience a great deal."
"Have you anticipated, then, that *your* reader may not be able to follow you at this point?"

In what possible sense is the audience (or the reader) the writer's? What habit of mind or belief could be said to justify that claim? Perhaps, it could be argued, we are not considering sediment here but a *conventional* expression, as in "my doctor" or "my lawyer." In those, however, possession is reciprocal: "my patient," "my client." With "my writer," the reciprocity sounds ridiculous:

"Have you thought about your writer?" said the first reader to the other.

"Yes. I've thought about my writer a great deal."

"Do you think, then, that your writer anticipated your not being able to follow at this point?"

If the first conversation seems sensible enough and the second not nearly, they do so because of our habit of considering audience as the writer's possession, a rather passive receptacle to be informed or reformed, though already determinate, well-formed, identifiable, and real.

Given the discussion in part I, we ought to be able to hypothesize that our notions have changed, away from this conception of audience as fixed and determinate and toward audience as fluid and indeterminate, away from audience as real and toward audience as figure or fiction. This hypothesis is worth exploring, because if audience is not fact but figuration, we must understand who or what configures it.

As Ede (1984, p. 141) notes, much of the earlier work on audience analysis (that dating to the early 1950s) was very much influenced by Aristotle. As a result, traditional audience analysis, according to Park (1986), takes "literally the idea of 'knowing' an audience as examining a group already assembled and describing any or all of the characteristics that those assembled may happen to have" (p. 480). These characteristics are demographic: age, sex, race, religion, political affiliation, education, income, and the like, characteristics often used in mass communication (e.g., advertising) to target various messages. Once these characteristics (or the traits and qualities Aristotle called character) are identified, the argument can be adjusted accordingly. Note that in this view, the audience doesn't sanction the discourse (because the audience doesn't "own" the discourse but "belongs" to the writer or speaker); rather, the writer originates the message and *then* adapts it.

To this demographic orientation was added another conception in the 1960s and 1970s, when reader-response critics like Booth (1961) and Iser (1978) distinguished between the actual reader and an *implied* (or *fictional* or *mock*) reader. Unlike the actual reader who processes the text and is external to it, the implied reader is a textual construct, an abstraction existing within the text that allows the actual reader to adopt a particular persona. In his influential article "The Writer's Audience is Always a Fiction," for example, Ong (1975) demonstrates how Hemingway's use of the definite article and the demonstrative pronoun offers a reader a specific role to play in *A Farewell to Arms,* that of boon companion.

The best-known combination of these two orientations is Ede and Lunsford's (1984) article "Audience Addressed/Audience Invoked," whose categories correspond to reader actual and reader implied. Their formulation, according to the authors, provides "the most complete understanding of audience [which] involves a synthesis of the perspectives . . . termed audience addressed, with its focus on the reader, and audience invoked, with its focus on the writer" (p. 167).

This "most complete understanding," however, is not a *synthesis* of the two orientations but a combination of them. The formulation does indeed bring both parties to the parlor, rather than leaving one at home, but there they sit, a concrete reality on the one hand and a textual construct on the other, albeit circumscribed by the rhetorical situation. What concretizes that reality, we may wish to ask, and what sanctions that construct, and in so doing authors and authorizes them both? To answer this question, we need to turn to K. Burke's (1961) often-quoted description:

> Imagine that you enter a parlor. You come late. When you arrive, others have long preceded you, and they are engaged in a heated discussion, a discussion too heated for them to pause and tell you exactly what it is about. In fact, the discussion had already begun long before any of them got there, so that no one present is qualified to retrace for you all the steps that had gone before. You listen for awhile, until you decide that you have caught the tenor of the argument; then you put in your oar. Someone answers; you answer him; another comes to your defense; another aligns himself against you, to either the embarrassment or gratification of your opponent, depending upon the quality of your ally's assistance. However, the discussion is interminable. The hour grows late, you must depart. And you do depart, with the discussion still vigorously in process. (p. 115)

Of this little narrative, we might say the following. First, to foreground *you* as rhetor addressing an audience is to ossify the flux of exchange, for within the scene of conversation there exist no speaker qua speaker and no audience qua audience, but speakers and interlocutors who become one another. Second, you did not invent your argument; its subject was given to you by the group (who neither originated nor will end it), as were the range of possible claims you could use and the genre to employ them. Were you to enter another parlor in, we can pretend, a house of parlors, the argument and possible claims and genre could be different, that is, locally determined. Third, as G. Clark (1990) writes in a similar context, the parlor conversation exists "neither to represent reality nor to transmit it, but to constitute it" (p. 1). In your absence, a conversation would have occurred, but that conversation would not have been the same. In the absence of all conversation, the parlor would exist, but it would be a different parlor.

None of this is intended to counter the claims of those who see audience as readers in the text or of those who see audience as concrete reality initiating discourse. I am real and so do you believe yourself to be. And in writing these words, I have fictionalized you, have created for you various roles my idea of you might play in reading them. You may choose to adopt those roles or not, as you fictionalize me, as to-gether our personae dance, perhaps convivially, perhaps otherwise. As dancer, you have some power and, as dancer, so do I. But around and interpenetrating us lies a matrix of social interaction that has con-strained and empowered us both. From that perspective we are authored and we authorize in turn. How can we know the dancers from the dance?[2]

Because it recognizes the existence of the dance, R. Miller and Heiman's (1985) *Strategic Selling* is a most powerful text on audience analysis. *Strategic Selling* demonstrates that, prior to their being *con-crete realities,* audiences (and, by implication, rhetorical situations) are organizational and hence social constructions, formulated as roles that are directly related to the way knowledge is constituted, stored, and communicated in the organization. The book focuses not at all on writing or speaking but on selling, in a future-shock world where "instability" (R. Miller and Heiman, 1985, p. 22) is the only constant and "corporate structures . . . are in constant flux" (p. 55). Given the rapidity of this change, the seller *first* needs to look, not for individual people (concrete realities) within the organization or for their titles (demographic information), but for roles, three of which exist in any complex sale.

The role of the Economic Buyer, as defined by R. Miller and Heiman, is to approve the sale. Because this person has direct access to the money and its discretionary use, he or she can manipulate the budget to "discover" or release unbudgeted funds. Although this role may be per-formed by a collectivity, one player within that group is generally the first among equals; thus there exists only one economic buyer in each sale. The role of the User Buyers (there are often several in each selling situation) is to judge the impact of the offered product or service on job performance. In the sale of group insurance to a very large company, for example, user buyers could include not only the employee benefits manager but also the personnel manager and the insureds themselves. The role of the Technical Buyers is not to decide who wins but who can play; they are the gatekeepers who employ quantifiable criteria (specifications, price, delivery time, and the like) to screen out the various sellers. Because of the instability and flux within the organization, "a given player in an account can shift roles quickly and unpredictably, even though the player's title and official function on the buying 'team' remains the same" (R. Miller and Heiman, 1985, pp. 71-72). In another selling situation, moreover, in the same division of the same company, the previous Economic Buyer may be a Technical or User Buyer, depending on "the dollar amount of the sale, business conditions, the buying firm's experience with [the seller and her company], the buying firm's experience with [the seller's product or service], and the expected organizational impact" (p. 74).

Only after the key buying roles are identified can the seller begin to respond to each individual's "receptivity to change" (p. 58), can begin, that is, to consider each as a concrete reality, and determine how "to serve their individual, subjective needs as well" (p. 61). Crucially, that subjectivity is not foregrounded; its possibilities are constrained by the community-generated role the individual plays in *that* situation. These roles define and limit the range of concerns to which a given individual is apt to attend in playing a role. The roles empower by providing a place in the narrative and constrain by delimiting the range of response.

Given that the organization or the group authorizes its audiences as well as the possible claims that can be emplotted in its possible narratives, we need to consider how the structure of organizations has been changing in response to the postmodern condition, for the organization or group, it appears thus far, defines the shape of the field on which practice can be played.[3]

III

The privileging of the local dimension should not be seen as running counter to changes in the last 30 years in the global economy, which has become increasingly integrated and which, considering events in the Eastern Bloc and the standardization of the European Community, will become increasingly more so. Rather, the global economic system, as Cooke (1990) writes, actually has no center; it is characterized instead as "a decentred space of flows rather than a clearly hierarchically structured space of production" (p. 141). Within this decentered space exists the individual corporation itself,[4] where much of practice is played and whose organization of production has become increasingly decentralized and heterogeneous. This movement contrasts sharply with the model of the earlier so-called Fordist corporation whose vertical integration demanded hierarchy and bureaucracy so that necessary control was exerted from raw materials to final production.

The Fordist corporation (and its hierarchical structure) was one way of organizing knowledge. Important relationships within the structure were vertical and limited, with labor a highly focused input valuable only on the level occupied by the individual worker. The root metaphor informing the structure was the machine, and the worker was a replaceable part. The clear relationship between part and function suggested that knowledge was, and its transmission could be, determinate and predictable. The structure was an efficient method of organizing knowledge and information when storage capacity was limited and when change was slow—that is, when quick retrieval was not a necessity.

Our notion of the corporation as Fordist is one reason why we tend to view corporations in general and the large corporation in particular as a monolithic and homogeneous discourse community.[5] The considerable diversity within the corporation was masked and muffled in the Fordist firm, which encoded a masterplot within the organization, "a kind of imperialism," according to Toffler (1990), "governing the company's diverse hidden 'colonies.' " "The struggle to rebuild on post-bureaucratic lines," he writes, "is partly a struggle to de-colonize the organization—to liberate these suppressed groupings" (p. 184). This decolonization has been under way for more than a decade. Disappearing are the hierarchical, rigid, homogeneous structures that characterized the Fordist corporation; replacing them are *flattened* structures that encourage flexibility, heterogeneity, and a ferment of critical discourse.[6] These structures mark a departure from the traditional discourse of scientific management that relied on rigid and hierarchical chains of command and reporting. Each, instead, recognizes the efficiency and performativity[7] of difference and encourages a greater diversity of discourse.

In a compelling analogy, Toffler (1990, pp. 174-176) notes the striking similarities between these organizational changes and the development of computers. In the bureaucracy, information was channeled to designated departments or cubbyholes very much the way information in the old mainframes was stored hierarchically in memory. Computational power was concentrated at the top levels of the firm, and the brains were in the mainframe; the bottom-level machines, the *dumb terminals,* were unintelligent. With the advent of the microcomputer, however, information and computing power were accessed at the desktops, more widely distributing data bases and processing power. With the invention of relational data bases, the bureaucratic cubbyhole system was seriously threatened because lower-level users for the first time had the power to add, subtract, and interrelate fields in innovative ways. The real breakthrough, however, was the hypermedia data base, in which fields are related not in two dimensions but in three. Because information can be layered by level of abstraction, what increases is not just the quantity of accessible knowledge but also the significantly enhanced capability to combine, recombine, and juxtapose information to create new knowledge.

Toffler's (1990) analogy marks the similarities between how we structure knowledge and organize people: "When knowledge was conceived of as specialized and hierarchical, businesses were designed to be specialized and hierarchical"—and vice versa (p. 178). The new organizational structures will be less like conventional data bases, which are good for accessing information about the known, and more like hypersystems, which are effective when searching for the uncertain. Internally, the flex (as in *flexible*) firm of the future will contain a mosaic of diverse structures. Externally, it will rely even more heavily than present-day organizations on strategic alliances and the outsourcing of products, services, and information—that is, on *vertical disintegration* and the *diseconomies of scale.* These relationships will not be rigid and predesignated, as in the old data bases and in the Fordist firm, but fluid and constantly changing. Competitive position will be determined not only by internal resources, as in the vertically integrated company, but by "the pattern of relationships with outside units" (Toffler, 1990, p. 187). Indeed, Toffler continues, if much of the firm's "value added derives from *relationships* in [this] mosaic system, then the value a firm produces and its own value comes, in part, from its continually changing *position* in the super-symbolic economy" (pp. 230-231).

The flex firm, then, will be organized relationally like a hypermedia data base, and its value will derive from constant reconfiguration within the system. For the new fields of play, the new village greens, are like data fields, and they become meaningful only in the play of their differences, in their intertextuality, in their relationship to other fields.

"Data banks," as Lyotard (1984) writes, "are 'nature' for postmodern man" (p. 51).

IV

The collapse of the monolithic Fordist structure, like that of the totalizing metanarratives it resembles, unleashes the *petits récits* so long embedded in the larger story. The new organization, a mosaic of structures, voices and texts,[8] constantly changes not only internally but in its relationship to the decentered space of flows that characterizes the larger system of organized knowledge. The individual organization and the larger system are isomorphic with the new hypermedia data fields, and all of them are shaped by and shape in turn the way knowledge itself is constituted. Thoroughly affected, then, are the manner and access of storage as well as the available methods of transmission. And, hence, thoroughly affected is practice, for the practice of professional communication always involves the access, assemblage, and transmission of knowledge to interlocutors organized as a group. From a social perspective, such practice is always process, always yet another moment of conversation whose subjects and arguments, as in Burke's parlor scene, are locally determined, have neither beginning nor end, and inhabit the decentered space of the intertextual universe from which we speak.

Given the capabilities of the new information technologies, the materials of that conversation increasingly suggest Baudrillard's (1983) conception of the simulacrum, which develops as follows.[9] Assume that long ago, in the premodern period, a type of precious stone was extracted from the earth, worn for adornment, and coveted for its value. Later, in the modern period, technology was used to manufacture copies of the originals, whose value came to be derived, not in relation to the originals, but in the copies' superiority to other copies. By the time of the postmodern period, all the originals had been extracted; the copies' signifier had disappeared, and the copies, therefore, represented nothing beyond themselves. This is Baudrillard's notion of the simulacrum. Simulacra mark the postmodern condition of the hyperreal: Everything is as real as anything else, but everything is a nonreferential simulation. When a rock group performs "live," lip-syncing its prerecorded music, are we seeing (or hearing) the original or the copy? In the age of electronic reproduction, to alter slightly Benjamin's phrase, the notion of the origin becomes problematic.

Boilerplate offers a similar example. Several years ago, when I was teaching proposal writing to management consultants at a Big Six accounting firm, I had in my program a newly hired person who had been through the similar program I had conducted at his previous firm. In fact, I was then working for both companies, which, of course, are very much protective of their documents. Therefore I was careful not to use models extracted from one company's proposals in my programs for the other company.

The participant and I discussed a proposal that he had recently "originated" and that contained a particularly good benefits section more or less copied from another of his proposals written at the previous firm. In my subsequent discussion with others in the program at his current firm, I used that section as a model. When the participant later rejoined his former company and continued when appropriate to use the same benefits section, I also used it, without any impropriety, I believe, with participants in that firm. Needless to say, I have since come across quite a few benefits sections, written by numerous individuals at both companies, which appropriate much of the form and content of the "original." Assume that the original has been lost, a not unlikely case. Each new copy, then, refers not to its origin but self-referentially only to other copies.

This self-referential intertextual system forces us to rethink our ideas of creativity and of authorship. Imagination now becomes, not the process of forming visual images of objects or of imitating nature through symbols, but the "capacity to articulate what used to be separate" (Lyotard, 1984, p. 52). For in a world of copies, of simulations without origins, originality means assembling copies in a new way, like recombinant DNA, by taking existing bits and bytes of text and recombining them. Thus not only is the author already written, by the prescribed roles that the organization or group authorizes him to play, but his materials for discourse are already inscribed, in the intertextual system that allows him to speak. This so-called death of the author, however, doesn't mean his demise, only that the scribe is always already circumscribed. At risk here is not the author but what Kent calls elsewhere in this volume the master-narrative of objectivity, the notion that the Subject and the subjects of discourse lie outside language and the system of knowledge it enables.

Notes

1. Lyotard defines postmodernism as the incredulity toward master narratives.

2. Cf. Cooper (1986): "These attitudes, procedures, and arrangements make up a system of cultural norms which are, however, neither stable nor uniform throughout a culture. People move from group to group, bringing along with them different complexes of ideas, purposes, and norms, different ways of interacting, different interpersonal roles and textual forms. Writing, thus, is seen to be both constituted by and constitutive of these ever-changing systems, systems through which people relate as complete, social beings, rather than imagining each other as remote images: an author, an audience" (p. 373).

3. Because of space limitations, part III focuses on shape and not on tone and tonality (the sound and feel, the culture) of the organization. Cultural treatments of tone and tonality abound, many of them influenced by the work of Geertz and others who have popularized the "anthropological turn." See, for example, Deal and Kennedy (1982), Martin, Feldman, Hatch, & Sitkin (1983), Schall (1983), and Trice and Beyer (1984).

4. I am limiting the discussion below to the corporate environment, which is much more susceptible (than, say, university and governmental structures) to changes in the structure of knowledge.

5. Another reason has to do with the most visible structure of these organizations, especially the conglomerates. This structure is indeed hierarchical, marking the relationship between the parent and its divisions, but it really exists, according to Toffler (1990, p. 229), only for accounting purposes and belies the cultural heterogeneity of its parts. An analogous case is the European Community (EC), which regularizes currency and standardizes tariff and trade regulations, thus making more efficient the distribution of goods and services within the community. Despite the standardization allowing the EC to streamline its accounting function, the countries constituting this "conglomerate" are extraordinarily heterogeneous, and each country itself is composed of ever more heterogeneous structures.

6. Just-in-time distribution systems, total quality systems, and concurrent engineering all depend on such horizontal, as opposed to vertical, structures of organization. With concurrent engineering, for example, one kind of engineer might be charged to design a product; another, to build the equipment to produce it; and yet another, to develop the system to test it. In the past, each task would have occurred sequentially. This team approach is increasingly familiar to technical communicators. Many writers of documentation manuals, for example, now work closely with software writers and product engineers not just to understand the technical aspects of the product but to produce the manual itself.

7. According to Lyotard (1984), performativity (like paralogy) is a form of postmodern legitimation. Lyotard himself, however, does not subscribe to legitimation by performativity.

8. Given this diversity, we can see the importance of defining discourse communities, not only by attending to cultural factors but also by considering the new structural realities. According to Toffler (1990), "the shipping operation of [Company A] and the stock-intake operation of [Company B] form, in effect, a single organic unit—a key relationship. The fact that for accounting purposes, or for financial reasons, one is part of Company A and the other a part of Company B is increasingly divorced from the

productive reality. In fact, people in each of these departments may have more common interest in and loyalty to this relationship than to their own companies" (p. 229).

Thus cultural readings that view discourse communities as *only* company specific tell only part of the story. An analogue is the nation-state. In Eastern Bloc Czechoslavakia, for example, which for 40 years suppressed the Gypsies, it was easier (though misleading) to consider that group to have more common interest in and loyalty to Czechs than to Gypsies in other countries. Similarly with the Moldavians and other long-suppressed voices in the USSR. Totalitarianism, of course, is a master narrative.

9. Excellent discussions of Baudrillard's thought can be found in Wakefield (1990) and Kellner (1989); the latter book is less sympathetic, although more comprehensive.

INTERPRETATION

12

Gender Studies:
Implications for the Professional
Communication Classroom

MARY M. LAY

In career preparation, professional communication students must learn how gender roles affect communication behavior. Culturally and psychologically determined gender roles often influence language choice as well as interpersonal, leadership, and collaborative strategies. Whether writing a memo, preparing a speech, or assuming a group role, professional communicators who are aware of not only their own gender schema but also that of their audiences produce more effective communications. Instructors within the professional communication classroom should introduce students to the effects of gender roles. This discovery can free students from the limitations these roles place on their communication styles.

Moreover, because success in many classrooms and workplaces may be defined by the communication behaviors of the dominant or empowered sex, instructors should create pedagogical approaches that afford both sexes opportunity to influence these communities. By disclosing the cultural and social histories that students and instructors bring to classroom discourse, by admitting that the traditional classroom is not

a neutral place, instructors can confront and change classroom practices
that disadvantage students. In turn, students educated within this new
context of awareness may affect positively their future workplaces.

In this chapter, I discuss aspects of psychological and cultural gender
roles that pertain to communication and learning style, particularly that
of the female student whose values and learning modes may be neglected in traditional classrooms. To do so, I rely on the sociologists'
and learning theorists' studies of the female student's intellectual growth;
many of these studies apply findings of the object-relations school of
psychology. I then explain how reproduction and production educational theories relate to gender and the professional communication
classroom. Next I trace the dominant gender values and characteristics
of traditional classrooms (e.g., subject/object dyads and competitive discussion), and the professional communication classroom (e.g., adherence to scientific positivism). Finally, I suggest how instructors can "re-
see" the professional communication classroom by applying feminist
pedagogy to writing, reading, speaking, and group activities.

Gender Theory and Communication
or Learning Style

> Classroom discourse reveals meaning mutually created by teachers and
> students. It is never neutral, but is always situated in the context of a
> socially and historically defined present. (Weiler, 1988, pp. 128-129)

In acknowledging the social roles, including gender identity, that students bring to the classroom, Weiler's statement aligns with the production theory of education—meaning can be produced, challenged, or
changed within the classroom. The following section describes the
gender identities that men and women bring to classroom discourse and
the learning and communication styles of women that have been neglected in past studies of intellectual growth. The discussion also links
the production theory of education with the heuristic approach in professional communication pedagogy.

The Object-Relations School of
Psychological Gender Development

According to the object-relations school of psychological development, women and men who enter the classroom may have firmly estab-

lished communication styles based on their psychological gender identities. Chodorow (1978, pp. 93, 169) proposes that because women still have primary responsibility for parenting in our society, girls value connection with others while boys seek hierarchical distinction.

Gilligan (1982, p. 43) proposes that as a result many women develop an ethic of care and men an ethic of justice. In giving priority to others' needs and feelings, women suspect the abstract standards or hierarchical rankings that often govern the public arena, and they develop communication strategies that achieve and maintain connection.

Of course, the degree to which individual women value an ethic of care and men an ethic of justice varies as some individuals are more likely than others to process information according to society's gender schema (Bem, 1981, p. 355). However, regardless of how gender studies scholars explain individual differences *among* women and men, many propose common psychological differences *between* women and men that affect communication behaviors. These differences between women and men prevent the classroom from being a neutral place but situate it within a social context. The goal is to create appreciation for the communication strategies that women can offer the public arena.

Gender Identity of Female Students

Recent studies of how these gender differences affect intellectual growth are particularly important to the instructor. While Perry's (1968) and Kohlberg's (1981) stages of intellectual maturity are familiar to many educators, only in the last two decades have scholars proposed that these assumptions exclude women's experiences or consider the intellectual stance of many women "immature." Self-growth in women, studies assert, involves growth within relations rather than the separation, autonomy, and independence Perry and Kohlberg proposed in their studies of young males.

When Belenky, Clinchy, Goldberger, and Tarule (1986) apply Gilligan's theory of moral development to epistemological positions, they find that women may be "silent" or in positions of "received knowledge," "procedural knowledge," or "constructed knowledge" (p. 134). Women who achieve this last position value personal experience and relationships as well as the authorities they find in the public arena.

Similarly, Kaplan and Klein (1985) describe self-growth in women as including four planes: (1) a potential for entering into emphatic relationships, (2) tolerance for changes in relationships, (3) an ability to handle relational conflict and remain connected, and (4) the capacity to feel empowered by connection to others (p. 10). Certainly both men and women grow within relations. However, for decades intellectual maturity has been linked to "masculine" traits as defined by Kohlberg (1981): the discovery of universal ethical principles and the objective application of these principles to particular cases (pp. 409-412). Instructors and students must learn how these traditionally masculine and feminine principles affect communication behavior.

For example, when confronted with a narrative, female students might be "rooted," proposes Lyons (1990), "in the particulars of the situation and the relationships between people" (p. 69). Rather than attempting to solve problems, convince by arguing, and test for truth by finding consistency and logic, the female student is likely to seek understanding of situations, convince by identifying the characters' motives, and test for truth by measuring believability and intention.

This mode of learning stems from self-in-relation or *interdependence,* which for Stern (1990) includes distancing from and then returning to a relationship with an "enhanced capacity to love" (p. 82). For interdependent people, connection becomes a basis for maturity, much as the child's willingness to explore is based on assurance of nurturing support (Grumet, 1988, p. 27). Or, as Belenky et al. (1986) put it, "For women, confirmation and community are prerequisites rather than consequences of development" (p. 194). Interdependence, which may dictate the communication behaviors of the female student, insists that independence is most meaningful *within* relationships.

The female adolescent's tendency to look at the particulars of a case, to measure growth by enhanced relationships, and to prefer understanding rather than solutions can either be valued or suppressed in the classroom. Before turning to how this choice is manifested in the classroom context, the tension between reproductive and productive educational theory and prescriptive and heuristic teaching in the professional communication classroom must be clear. To a great extent, these educational modes embrace or reject the gender differences just explored.

Educational Theory and Gender Studies

The reproduction theory of education states that a society maintains its values, including definitions of appropriate gender behavior, through its classrooms. Schools then, according to Weiler (1988), "legitimate certain groups through variable access to knowledge and the use of language," and successful students may be intelligent or simply have ready access to what is socially valued (pp. 9-10). On the other hand, production education theorists study the ways individuals and groups "assert their own experience" to produce meaning in the classroom, even if this production entails confrontation, resistance, and change (p. 11). If women indeed value self-in-relation and an interdependent mode of learning, production theorists would ask how women can assert these preferences in the classroom.

Reproduction educational theory has much in common with the prescriptive approach to teaching business and technical communication, and production educational theory is similar to the heuristic approach. Defined recently by technical communication scholars J. Mitchell and Smith (1989), the prescriptive approach is "recognized as the more efficient way of training students to perform in predictable situations while the heuristic approach is more suitable for educating students to perform in a changing world" (p. 115). According to Mitchell, the function of technical communication is to transfer information, made easier if the reader is already familiar with the communication format. Smith proposes that versatility may be more important than knowledge of standard communication practices (J. Mitchell & Smith, 1989, p. 122). C. R. Miller (1989) would take this versatility one step farther: Technical communication is a matter of "arguing in a prudent way toward the good of the community rather than of constructing texts" (p. 23). For Miller, the central purpose of the communication is not to transmit information but to improve life.

Within the prescriptive approach to technical professional communication, reproduction rather than possible change in standard practices is of primary concern. If those standard practices disadvantage or exclude one gender, then the prescriptive approach to teaching seems to perpetuate rather than challenge that disadvantage or exclusion.

Gender Theory and the Classroom

American education does profound, lasting, psychological damage to many of its female students. Giving girls and women a woefully incomplete history of their sex, offering them literary texts considered canonical in part because of their projection of male fantasies hostile to women, insinuating that there are some subjects and some careers that are inappropriate for

women, allowing males to dominate classroom interaction, responding to a
woman's classroom comments with sexual harassment. (Smithson, 1990, p. 5)

The behaviors that Smithson indicts stem in part from associating
masculinity with rationality and objectivity, and this association appears in our professional communication classes in the form of scientific positivism. Such dichotomies not only exclude women's experiences from study but also prevent women from feeling comfortable in
classrooms that equate success with competition rather than cooperation. This section of the chapter explores the impact of gender dichotomy in the traditional and the professional communication classroom.

The Traditional Classroom

Gender as well as education scholars often comment on the dominant
masculine character of the traditional classroom. For example, Grumet
(1988) describes the classroom as "oriented toward a subject/object dyad"
(p. 22). Students master studied objects and by this mastery distinguish
themselves from the object. Gender dichotomy that privileges masculinity assigns to femininity the *other,* the *object,* or the *abnormal.* This
masculine epistemology, according to Bezucha (1985), comes from "the
separation of the public, the impersonal, and the objective, on one hand,
from the private, the personal, and the subjective, on the other" (p. 81).
The world, as studied in the classroom, is thus separated into contrasts:
the mind contrasted to the body, culture to nature, thought to feeling,
competition to cooperation, male to female. Masculine characteristics
and the resulting gender dichotomy can become so exaggerated that the
classroom establishes "success norms that favor males" (Grumet, 1988,
p. 45). To succeed, female students must adopt a style alien to their
gender identity.
    The style preferred by a good many male students creates a classroom
that is competitive, in which speech is aggressive and success is based
on individual distinction. Again, the experience of the male becomes
the norm for setting and testing abstractions, which, according to R.
Hall and Sandler (1982), makes it difficult for male students to "perceive women students as full peers, to work with them in collaborative
learning situations, and to offer them informal support as colleagues"
(p. 3). Instructors call on women less, ask them less probing questions,
and interrupt them more (Maher, 1985, p. 31).

Competition in the classroom may be threatening to females as it seems to sever connections. As a result, female students who do rise to the top may become depressed because of the "basic contradiction between the heavy pressure for individual, competitive achievement and women's motivations for action within a relational context" (Kaplan & Klein, 1985, p. 8). Conflict for many women is acceptable only if one can engage in it without "losing touch with the more basic affirming aspects" of connections and relationships (Kaplan & Klein, 1985, p. 5; see also Tannen, 1990, p. 149). To many women students, conflict and competition threaten to distinguish and so to sever relationships.

Often masculine values—the mind, culture, thought, and competition—thus are privileged by hierarchical ranking. Women may view hierarchies in general as "peripheral" (Noddings, 1984, p. 2). McIntosh (1985) describes a woman's reaction to success within hierarchical systems as "feeling like a fraud" (p. 1). She explains that a woman frequently begins a public presentation with what appears to be an apology rather than an assertion of authority. But this opening does not reflect insecurity as much as suspicion of the hierarchies that have disadvantaged her.

Finally, the positivism and rationality that characterize the classroom exclude women's experience. Students are taught that truth is discoverable through experimentation by the rational seeker. In a society that defines rationality as a masculine trait, women students may have little confidence that they can discover that truth, and in remaining convinced that there is a "truth" to be discovered, they fail to question their exclusion. Unfortunately, the evidence to support these truths seldom comes from women's experiences, from the private sphere, the personal relationship, the family, or the female body.

The Traditional Professional Communication Classroom

Because of its association with science and technology, the professional communication classroom may perpetuate dualism. When considered the objective transfer of data through standard formats, professional communication carries the remnants of scientific positivism, the belief that there is one truth, that it can be discovered using the proper methods, and that truth can then be transferred through communication (C. R. Miller, 1979). Although Kuhn (1970) challenged this myth of scientific objectivity, gender scholars, such as Keller (1985), propose that science still attempts to organize the world into dichotomies. And, in asserting that one is either objective or subjective, allied with nature or culture, governed by emotions or rational thinking, these dichotomies privilege what society labels as masculinity.

These biases remain even though most scientists and engineers admit that social values drive the scientific community. Indeed, Hacker (1990) discovered that engineering faculty ranked their fields according to association with pure or "hard" sciences. Electrical engineering was considered the "cleanest, hardest, most scientific" while the social sciences were dismissed as "soft, inaccurate, lacking in rigor, unpredictable, amorphous," in general "womanly" (pp. 116-117). Science, of course, has a long history of rejecting the feminine. In dealing with the *Woman Problem,* past scientists have measured skulls, weighed brains, and decided that too much education atrophies the reproductive organs. Present science seems somewhat bound to this legacy, as many girls lack confidence in math and science. A study of how science has dealt with women reveals just how much science is culturally and socially determined. (See Trecker, 1974, for a complete review of science's treatment of women.)

Professional communication, especially technical communication, sprang out of this scientific culture. Fortunately, the social values within science are now being seen as "collective choices . . . about what to objectify, what to predict, what to control, and what to know" (Bleich, 1990, p. 238). Science thus becomes not truth but choice, and the role of professional communication is to reveal and convey those choices. Professional communication instructors must teach students how to create communications that will best help readers understand and respond to those choices.

How far has professional communication come in changing this image? Bleich despairs that academic discourse has not even made the progress that science has in self-reflection and declaration of values. According to Bleich (1989), academic discourse and expository prose taught by academics deliberately exclude experience and give precedence to an ideal of "pure truth" (p. 19). But certainly within the heuristic approach—free from prescriptive forms—professional communication has made some progress. As D. Sullivan (1990) says, a professional communication course must "at once teach the discourse appropriate for the technological world *and* make students aware of the values embedded in such discourse and the dehumanizing effects of it" (p. 379).

Professional communication instructors can begin to meet these challenges by studying and adopting the characteristics of the feminist classroom, a classroom that changes how students have traditionally been encouraged to learn, relate to the instructor and each other, speak, write, and read. Some of these practices involve radical modifications; some are already in place within the heuristic classroom.

Re-Seeing the Professional Communication Classroom

> If we give students a double vision of social reality, I think they can learn both the language of power, which we use standing at the podium and delivering those straight sentences, and the language of social change, which suggests alternative visions of how to use power. (McIntosh, 1985, p. 9)

With its emphasis on collaboration, cases, and audience and language analysis, the professional communication classroom provides a setting for not only teaching students *language of power* but also helping them appreciate multiplicity, the authority of personal experience, and collaboration. In this classroom, the goal of which was once thought to be prescription of formats, lies the means to teach the *language of social change*.

The Feminist Classroom

The feminist classroom, most gender studies scholars agree, displays characteristics that include collaborative learning, appreciation of multiple perspectives, knowledge based on lived and personal experience, shared authority, and self-disclosure of feelings. Instructors in the feminist classroom primarily engage in production pedagogy, which allows students to challenge or resist social and ideological forces.

Within this classroom, rather than claiming authority, the teacher acknowledges personal background and gender identity. The teacher challenges sexist notions of appropriate behavior or work for women and encourages students to test theory against personal experience or derive theory from that experience. Moreover, the teacher allows students to challenge and replace theories that do not stand the test of experience. In the feminist classroom, students learn to value the "female" way of knowing, writing, or speaking. However, because the teacher also wants students to succeed in new social or discourse communities, students are taught the language of power; thus they receive a double vision of social reality.

Rejecting objectivity, locating authority within personal experience, and working for change within existing social institutions are themes that can inform the professional communication classroom. Professional communication instructors teach students to analyze their audiences, situations, and contexts. Acknowledging how multiple perspectives *created* the information that is passed to that audience is simply one step in the analysis.

The professional communication classroom can be feminist, for, as Schniedewind (1983) states, "Feminism is taught through process as well as formal content" (p. 271). Belenky et al. (1986) assert that all educators can help women students develop "if they emphasize connection over separation, understanding and acceptance over assessment, and collaboration over debate" (p. 229). This process should not be alien to the professional communication instructor who engages in heuristic pedagogy or who has landed on Bruffee's (1983, 1984, 1986) side of the process versus product debate in composition. How then can specific classroom exercises and assignments in the professional communication classroom reflect feminist pedagogy? The suggestions in the following section were tried in my junior/senior level business communication course. The course text was Tebeaux's (1990a) heuristically based *Design of Business Communication*. The suggestions exemplify ways of integrating gender concerns in the classroom.

The Professional Communication Classroom

Learning

In the professional communication classroom, students often explore case studies, engage in collaborative writing, and critique each other's writing within groups. This setting then naturally invites collaborative or *connected* learning. (See how the teachers at Emma Willard School tried this kind of connected learning in math, philosophy, and psychology classrooms, Gilligan, Lyons, & Hanmer, 1990, pp. 286-313.)

If, within case studies that lead to writing assignments, students are asked not only to convey information but also to maintain a relationship, personal experience takes on additional value. For example, student writers often must reject a reader's request while preserving "goodwill." In one of Tebeaux's cases, the writer must inform a respected authority who has suffered a stroke that he cannot speak at a conference that he helped create. Students can discover how to reject him tactfully if they recall their own experiences with rejection. In fact, Tebeaux (1990b) herself, in a study of MBA students' responses to this case, found that "people" experience in the workplace helps students handle this situation (pp. 29-34).

Students also learn to trust their own experiences by becoming their own audiences. For example, in another of Tebeaux's cases, the writer must request new computer equipment from a supervisor who has turned down all recent requests. Students can understand the situation further by exchanging their requests and in the role of the supervisor reject the request. Because the students wrote the original request, they can better create a tactful rejection. Also, such a reversal validates the students' experiences and returns authority to them.

Teaching

Feminist pedagogy calls for instructors to reveal their own experiences and values, to discuss how gender may determine these traits, and to diminish any distance created by their authority. While some instructors are comfortable with self-disclosure, I found that revealing the values of the course textbook writer can just as easily demonstrate that authorities are not neutral and objective.

For example, before their first reading assignment in the Tebeaux text, my students read an article written by Tebeaux (1988) in which she presents her position on professional communication (p. 17). Upon reading the article, students discover their textbook writer's professional stance and analyze how she views her audience and purpose. Since most textbook writers also publish in scholarly journals, such sources are readily available. Because students discover their textbook writer's professional attitudes, they read the text critically. They perceive themselves as the audience Tebeaux wants to motivate, and she becomes a less distant authority.

Reading and Writing

The language students read and write also reflects gender values. While culture creates language, language in turn reproduces culture. These concepts, promoted by Sapir (1921) and expanded by Whorf (1956), have been adapted by those promoting gender equality. In the professional communication classroom, gender studies inform language use, genres of writing and reading, and gender preference in writing style.

Language and gender scholars now promote nonsexist language. However, D. Hall and Nelson (1990) found that while professional communication textbooks encouraged students to use nonsexist language, little practice was offered (p. 73). If, rather than simply requiring nonsexist language, instructors also discuss the reasons for such change, students become more aware of gender roles. For example, instructors can explain that *chair* is preferred to *chairman* because, according to Treichler and Frank (1989), *man* "no longer functions generically for many people" (p. 197; see also, C. Miller & Swift, 1980, pp. 25-26; Tannen, 1990, p. 243). People visualize a male when they read *chairman* and cannot see a woman in the job or see a woman as an exception.

Moreover, instructors should discuss with students the special power of professional communication to change gender roles. For example, Dell (1990) explains that because readers often believe that technical writing is "factual, truthful, and precise," they may have " 'turned off' many of the filtering screens they normally use in judging writing" (pp. 250-251). Using sexist language in professional writing may reproduce sexual discrimination more readily than other genres; on the other hand, freeing technical writing from sexist language helps promote sexual equality.

Gender scholars are just beginning to ask whether men and women have different writing styles (e.g., Flynn, 1988, p. 117). Unfortunately, some studies reveal that the traditional female voice is graded more harshly even in women writers than the traditional male voice (Barnes, 1990, p. 152). Tebeaux (1990b) finds that while beginning female writers seem to have more tact than beginning male students, male students develop equal ability with work experience or instruction. However, studies done by Smeltzer and Werbel (1986) as well as Sterkel (1988) discern no measurable difference between women's and men's styles in business communication.

On a more challenging level, the traditional genres that instructors prescribe may reproduce sexual inequality. According to Daumer and Runzo (1987), students learn a style "whose distinctive features are detachment from others, suppression of emotion, a 'logical'—i.e. hierarchical—organization, 'appropriate' topic and word choice, persuasive strategies and reliance on rules" (p. 52; see also Bleich, 1989, p. 19). Freeing stylistic characteristics from gender association may create communications that more effectively convey information and engage the reader. For example, in the computer industry, selected customers test whether instructions work well for them (e.g., Grice, 1991). Customers' education and experience determine the document's metaphors, examples, and patterns of organization. If writers envision only male users, feminine metaphors and experiences are neglected. Freeing language from traditionally masculine genres increases students' choices of language and organizational patterns.

Finally, whenever reading a professional communication, students should be encouraged to look for gender bias. Foss (1989) suggests students ask about all documents: "Does the artifact describe how the world looks and feels to women or men or both?" "How are femininity and masculinity depicted in the rhetorical artifact?" and "What does the rhetorical artifact suggest are the behaviors, concerns, issues, values, qualities, and communication patterns of women and men apart from the society's definition of gender?" (p. 1). Students can ask these questions about their own writing and about the examples, models, and cases they encounter in their textbooks.

Speaking

Gender identity may also affect the oral presentation and classroom discussion styles of professional communication students. In the traditional classroom, students may be taught "little professor talk," to assert their expertise, to refute others' comments, and to separate their ideas from others'. However, Kramarae and Treichler (1990) found that women are most interested in classroom talk that supports friends, in "shared discussion" that includes the experiences of other students, while men prefer teachers who "organize most of the class content through lectures and who encourage questions and comments from individual students" (pp. 54-55). Instructors who value connective discussion as well as individual participation not only support their female students but also might encourage all students to become more versatile in their discussion style. (See Aires, 1987, for a complete review of verbal, nonverbal, and topic selection of female and male students.)

Group Activities

Although students in the professional communication classroom may engage more in group than in open discussion, scholars propose that males and females may behave differently in small groups. For example, Baird

(1989) found that in small groups males are more competitive, more likely to assume leadership roles, and more task-oriented than females; females are more willing to self-disclose, more cooperative, and more interested in interpersonal dynamics (p. 192). Lay (1989) found, moreover, that females may view conflict within the group threatening, whereas males may enjoy conflict and debate (pp. 20-21). Unless understood, these differences may disrupt the group.

The kinds of talk in which group members engage also may reflect gender difference. Duin, Jorn, and DeBower (1991, pp. 159-160) and Van Pelt and Gillam (1991, pp. 194-195) discovered that collaborative writing groups must engage in *off-task* or *social* talk. Tannen (1990) calls this *troubles* or *rapport* talk—the talk many females use to establish connections (p. 58). Because this talk promotes group bonding, all students should learn to value it as much as on-task talk, as a group must engage in community building so that it can weather times of conflict. (For a description of the necessity of conflict, see Burnett, this volume, and Karis, 1989.)

Students also need to understand the gender-based leadership styles within their groups. Baird and Bradley (1979) found that the "participative" style of leadership that involves all group members in decision making, that invites openness and trust, typified popular and effective female managers (p. 109). In participative leadership, listening becomes a leadership strategy. Young female leaders interviewed by Lyons, Saltonstall, and Hanmer (1990) define listening as "an activity of competence—following the reasoning of people, weighing ideas against other ideas, including one's own, and developing new options" (p. 186). Leadership for many female students may then be interdependent rather than autonomous, participatory rather than hierarchical. (See Helgesen, 1990, for interviews with women executives who exercise this style.)

Scholars are just beginning to study the collaborative writing strategies of men and women. In identifying two collaborative modes —the hierarchical and the dialogic—Ede and Lunsford (1990) found more women than men seem to engage in the dialogic mode in which the "group effort is seen as an essential part of [knowledge] production" instead of "a means of individual satisfaction within the group" (p. 133). To ensure a cooperative classroom, students should help define success even with individual assignments. For example, on the day students hand in a completed assignment, I have them share with me their problems with and assumptions about a case. They feel empowered when they let me know before I grade an assignment what they experienced in completing it. In essence, they tell me what to look for when I evaluate their work.

Because students in professional communication classes engage in a large amount of group work, instructors must help students appreciate the many gender differences that may emerge within these settings.

Gender Perspectives in the
Professional Communication Classroom

As social and familial structures change, students may enter our classrooms with communication behaviors no longer limited by gender roles. Until that time, if we welcome in our classrooms women's communication strategies, we may free both males and females from their traditional roles. In the professional communication classroom in particular, the heuristic approach to teaching encourages students to produce knowledge and to propose change within their discourse communities. Also, as instructors we need to question distinction through competition and conflict, hierarchical ranking, and mythical objectivity and consider how collaboration, multiple perspectives, shared authority, and personal experiences can enhance our classrooms. We must allow gender theory to inform the way professional communication students learn, read, write, speak, and work in collaborative groups, as well as the way we present ourselves.

In addition, gender studies suggest that gender roles affect readings and assignments. The professional communication classroom curriculum should be subjected to Foss's (1989) questions, modified as follows: Do the course content and assignments describe how the world looks and feels to women or men or both? How are femininity and masculinity depicted in the course content and assignments? Do the course content and assignments suggest behaviors, concerns, issues, values, qualities, and communication patterns that are free from society's definition of gender?

13

The Group Writing Task:
A Schema for Collaborative
Assignment Making

MEG MORGAN

In its most basic form, social constructionist theory maintains that groups rather than individuals create reality. The assumption for teaching behind this simple—yet complex—statement is that if we want our students to learn whatever we are teaching, we should allow them to have an interactive, social experience so they can create their own realities. The practical outcome of such thinking is the movement toward using groups in classrooms, and most specifically for this book, in professional writing classrooms. This movement toward using groups has, over the past few years, produced many wonderful, creative, and challenging collaborative writing assignments.

While I believe that such a shift is for the better, because it emphasizes sharing, cooperating, and including rather than excluding, I want this chapter to serve a cautionary function: Although grand theories—social constructionist, for example—influence the way we think about reality, by themselves such theories will not inform our practice without what Merton (1968) calls theories of the middle range, or theories that "lie between the minor but necessary working hypotheses that evolve in abundance during day-to-day research and the all-inclusive systematic efforts to develop a unified theory" (p. 39).

Middle-range theories evolve from empirical research and in turn inform further research or in the case of assignment making, the practice of pedagogy. Without middle-range theories, we are literally creating in a vacuum, using only the lens of the *unified* theory to inform what we do. In composition studies, for example, Flower and Hayes (1981) developed their model of the composing process from the unified or grand theory of cognitive psychology to aid compositionists in their work. Deriving from both a unified theory and empirical research, their model, which I would call a theory of the middle range, provides bases for further empirical study and for teaching.

When we design writing assignments based only on the unified theory, we design with no parameters, and because we have no empirically derived or empirically tested theories on which to design, we often make mistakes, perpetuate those mistakes, and inevitably fall back on a pedagogy based on anecdote. The pedagogy of collaborative writing is still fundamentally anecdotal because no middle-range theories bridge social constructionist theory and collaborative classroom practice.

The major purpose of this chapter is to create a schema—a middle-range theory—to help teachers make more informed decisions in designing group assignments for professional writing classes. The schema is developed from research in small-group communication and collaborative writing. The discussion focuses only on *shared document* collaboration (Allen, Atkinson, Morgan, Moore, & Snow, 1987, p. 84), documents written by groups, not documents written by individuals and critiqued by groups. Before presenting the schema, the chapter reviews how tasks have been classified in small-group and collaborative writing research. Finally, this chapter argues that some collaborative writing tasks we use are redundant and inappropriate for groups, and that students need tasks that appropriately test a variety of group skills in a variety of group situations.

Classifications of Tasks

Middle-range theories emerge from the interaction between grand theories and empirical research. I used theories from empirical research on task types in small groups and on collaborative writing to develop my schema. Two questions guided my inquiry:

1. What are appropriate kinds of group tasks for collaborative writing?
2. What kinds of activities are appropriate for writing in groups?

What Are Appropriate Kinds of Group Tasks for Collaborative Writing?

Small-group research provides help with classifying group tasks. Although several typologies in small-group research classify tasks, I have chosen Steiner's typology because his types seem closest to the kinds of assignments that appear in published accounts of collaborative writing. Steiner (1972) divides tasks into three sets of polar categories: unitary/divisible, maximizing/optimizing, permitted/prescribed (pp. 15-17). These three categories interact with each other, so a complete picture of any task would examine how the task fits into all categories, for example, how a task is unitary, optimizing, and prescribed, or divisible, maximizing, and prescribed. For clarity, the following discussion separates these categories.

Unitary/Divisible

A task is unitary when it cannot be divided into subtasks performed by different people or when no practical purpose is served by dividing the task. A nonwriting unitary task might be counting gumballs in a fishbowl in order to win a prize. Although one might argue that this counting task could be subdivided (count all the red balls, then all the green balls, etc.), subdividing seems unnecessarily cumbersome and under most conditions serves no practical purpose. A divisible task, on the other hand, can easily be divided into subtasks. Even small tasks, like salad making, can be divisible: First make the dressing, then chop the vegetables.

Groups can perform both unitary and divisible tasks. For example, a group can decide how many gumballs the fishbowl contains by collecting guesses and then averaging them for one answer. A group can participate in salad making by allocating subtasks to several individuals.

Maximizing/Optimizing

Groups can also perform maximizing and optimizing tasks. Maximizing tasks are those requiring that individuals or groups do as much as possible in as little time as possible, with success usually measured in quantitative terms. Optimizing tasks, by contrast, emphasize the quality of the outcome. In an educational setting, the absolute dichotomy set up by Steiner (1972) gets blurred. We often want our students, for example, to do the most in the shortest amount of time, but we also want quality —correct answers. As teachers, although we design both maximizing and optimizing tasks for students, certainly the balance is toward optimizing.

Permitted/Prescribed

Steiner divides tasks into those that a group is permitted to accomplish and those that must be accomplished by a group. In a classroom context, teachers seldom allow students this option: A writing assignment usually is or is not to be done in groups. Four subcategories—disjunctive, conjunctive, additive, and discretionary—describe, in a group situation, the choices permitted or prescribed to a group to complete a task. Only disjunctive and conjunctive tasks are directly applicable to my discussion. Additive tasks (those that absolutely require a group and 100% performance by every member to complete the task) don't really seem to exist in shared document collaboration; discretionary tasks (those in which task division is the group's decision, not imposed by the task or some outside authority) seldom occur in classrooms because teachers do not give that authority to students. For example, we usually would not permit only one student to take total responsibility for a group assignment. Disjunctive and conjunctive tasks are pertinent to this discussion.

Disjunctive. A disjunctive task requires only one competent person to complete the task and, in fact, may not even require group participation. Creating a group to complete the task, however, increases the chance of finding a competent person. A disjunctive task thus is one where chances for successful completion are enhanced by adding members to the group. The gumball example is a disjunctive task if only one person is needed (permitted or required) to come up with the correct answer. If the gumball directions say, "As a group, look at the number of gumballs and decide on a correct estimate," then only one group member with a particularly keen eye and an extraordinary streak of luck may be needed to come up with a winning guess. Solving a crossword puzzle is also a disjunctive task: One person could complete the entire puzzle but it may get completed faster if one person does all the down words, another all the across words and a third staffs the dictionary. However, increasing the size of the group may not be effective past a point "beyond which additional members have no added effect, either because the most competent member cannot be readily recognized or because the group is already large enough to insure inclusion of at least one member capable of completing the task efficiently" (Shaw, 1981, p. 175).

In small-group research, disjunctive tasks are often maximizing tasks in which quantity or time determine the successful completion of the task. Thus a group (or even a single member of a group) that produces as many answers to a crossword puzzle as possible in as short a time as possible is the most successful.

Conjunctive. A conjunctive task is one in which all members must be involved to complete the task. The gumball example becomes a conjunctive task if all members *must* participate; for example, all members must submit a guess, the guesses are averaged, and the average is submitted as the correct answer. Because participation of *all* members is *required,* one incompetent member will affect the group's performance. Adding members to a conjunctive task thus decreases the chances of successful completion, because it increases the chances of including an incompetent member. A missing or really wacko guess will throw off the average of the rest of the members. With a unitary task, if one person does not complete his or her part the task cannot be successfully completed. A divisible task, such as a crossword puzzle, could be a conjunctive task if the specialties of individual group members are required for completion.

In addition to small-group research, research in collaborative writing also provides a typology to help with the design of collaborative writing tasks. This typology developed by Allen et al. (1987, pp. 85-87) classifies types of writing groups (not tasks) in which group performance provides a clear advantage for the individuals or the organization. This typology is composed of three different kinds of groups: labor-intensive, specialization, and synthesis.

A labor-intensive group is formed to complete a task too large for any one person to complete in the allotted time, although the task can be completed by one person given enough time. In addition, any member of a labor-intensive group can perform any of the subtasks of the group or exchange tasks with any other member. In a specialization group, all members are necessary because successful completion of the task requires the expertise of each group member. Although a synthesis group forms to write a document, members must also meld divergent points of view. The outcome is twofold: a reconciliation of differences and a document that reflects group consensus. A synthesis group document need not be group written; the group's purpose may be to defuse (and diffuse) the sensitive nature of a situation or document.

What Kinds of Activities Are Appropriate for Writing in Groups?

Tasks can also be typed by "the kinds of cognitive 'matters' with which a group works" (Hackman, 1968, p. 164). Hackman developed three categories of task types based on content: production, discussion, and problem-solving tasks. Interestingly, Hackman empirically tested the typology using writing tasks.

Production Tasks. Production tasks require a group to produce "ideas, images, or arrangements" which often result in stories or descriptions (Hackman, 1968, p. 164). Groups engaged in production tasks often are more concerned than other groups with the quality and length of the group product. They also tend to emphasize the product—the outcome—while groups working on other tasks emphasize the process.

Discussion Tasks. Discussion tasks require "a discussion of values or issues, usually with a requirement for group consensus (Hackman, 1968, p. 164). A group engaged in a discussion task goes through a process of evaluation during which the group comes to a judgment on a single issue or problem. Because the group must reach this judgment through consensus, the product is deemphasized and the interaction among group members necessary to achieve judgment assumes primacy.

Problem-Solving Tasks. Groups engaged in problem-solving tasks must find solutions to specified, often practical, problems: writing a set of directions, judging the feasibility of a new procedure. Much of the work in problem-solving groups is planning and coordinating; although groups may need to judge, discussion and consensus building are not the predominant characteristics of problem-solving groups. Hackman and Vidmar (1970, p. 50) suggest that groups do not experience the "interpersonal tensions" on problem-solving tasks that they experience on production tasks for two reasons: Members of problem-solving groups do not engage in individualistic and divergent thinking characteristic

of production groups and members may have more experience working
in problem-solving than in production groups and are thus used to the
kinds of interactions that accompany such work.

The three typologies discussed above (Steiner, 1972; Allen et al.,
1987; Hackman, 1968) create the theoretical framework for the rest of
this chapter in which I present a schema that conceptualizes collaborative writing tasks.

A Schema for Group Writing Tasks

The schema I present in the following discussion merges Hackman's
(1968) content-based typology (production, discussion, and problem solving) with Steiner's (1972) typology of group tasks and Allen et al.'s
(1987) typology of groups to form a more complete typology and one
that functions as a middle-range theory for collaborative writing tasks.
After I present the schema, I describe and illustrate these task types,
drawing on published collaborative writing assignments for professional writing classes. The four task types I develop are *individual, melding, aggregate,* and *interdependent* (Table 13.1).

Individual Tasks

Individual tasks are unitary tasks that cannot reasonably be broken
into constituent parts and may be best accomplished by one person.
As much as possible, these tasks have a tendency to be maximizing,
where writers are rewarded for swift and timely completion and for
length or amount. Because of these characteristics, I suggest later that,
in fact, individual tasks are usually inappropriate for shared document
collaboration.

Individual tasks do not suggest any particular area of content according to Hackman's typology, although clearly they could be either production or problem-solving tasks, but not discussion tasks in which all
members must participate. However, the creation of images, ideas,
stories, or imaginary descriptions appropriate for Hackman's tasks
are often inappropriate writing tasks in professional writing classes.
Instead, assignments in professional writing classes may be product
oriented in two ways. First, although they do not generate images such
as stories or descriptions, they may generate images such as a graph or

Table 13.1 Schema for Collaborative Writing Task Types

Individual Task	Melding Task	Aggregate Task	Interdependent Task
Cannot be divided into parts	Cannot be divided into parts	Can be divided into parts	Can be divided into parts
Needs one person to complete	Requires group to complete	Needs one person to complete	Requires group to complete
Production or problem-solving task	Discussion task	Problem-solving task	Discussion or problem-solving task

chart. Designing visuals, especially by hand, might best be done by a single person. Second, the task may be so routine or familiar to the writer that little attention to the process is required. Such tasks might include simple memos or letters.

Problem-solving tasks also might provide a content area appropriate for individual tasks. Problem-solving tasks are high in "action orientation," produce little tension within the group, and allow group members to feel "less inhibited," less competitive, and more efficient (Hackman & Vidmar, 1970, p. 44).

Many teachers do not make an individual task collaborative. However, in Bosley's (1989) survey of professional writing teachers, more than 56% of her respondents report letters as their most frequent collaborative assignment (p. 110). In many letter writing assignments, the task cannot be divided and can be done by one competent person, so collaboration offers little advantage. On the other hand, using an individual task with a problem-solving orientation, such as asking a group to write a description of a simple mechanism within tight time limits, might be a way to introduce students to relatively stress-free group work.

Melding Tasks

Melding assignments cannot be broken down into subtasks, and all group members are needed to accomplish the task: Steiner's (1972) unitary/conjunctive tasks. Hackman's (1968) discussion tasks also typify melding tasks: The content often deals with issues on which group members' opinions diverge and the task brings all opinions together, requiring each member's input. Melding tasks also resemble those undertaken by the synthesis groups in the Allen et al. (1987) typology, that is, tasks that were small but the content of which generated dissension or disagreement. Melding tasks are clearly optimizing tasks—tasks that require high quality, not quantity, of output.

Melding tasks require sophisticated interpersonal, consensus-building, and decision-making skills. They often do not have clear goals from the start; goals emerge or are constructed through discussion and sharing. And even after the group has cohered and the document is written, no guarantees exist that the document or the solution it advocates will be acceptable to all group members or to the larger organization that this group may represent.

Few melding collaborative writing tasks exist in the literature of collaborative writing in professional writing classes. Beard (1990) uses an exercise in his business writing classes that involves a four-person group (pp. 65-67). Three of the four members must learn rules of a game and devise a strategy for winning the game; they must then write a memo describing the game rules and the strategy, and hand it to a fourth group member who was not part of the initial group sessions. The fourth member, using only the memo, must figure out the game rules and the winning strategy to help the group win the game. In this exercise, group members must agree on a strategy through consensus and must enact that strategy in order to win; no single member can win the game. Beard notes that there is no one solution to winning the game, and if students don't cooperate both as they play the game and as they write the instructions, the chances of winning are minimal.

Aggregate Tasks

An aggregate task is a divisible/disjunctive task, one that any one group member can complete. Allen et al.'s labor-intensive group performs an aggregate task because, although the task can be completed by one person, the several members of a group can accomplish the task in less time than an individual. Hypothetically, because the task can be divided into parts, group members would not even have to meet to complete it. Realistically, subparts are usually interdependent in even the most divisible tasks.

Hackman's typology suggests that only problem-solving tasks can be accommodated within this type: Production tasks are often not divisible and discussion tasks are not disjunctive. Like other disjunctive tasks, adding more people up to a point increases the likelihood of finding a person who can successfully complete the task.

An aggregate task often requires a number of operations that allows group members to take on different subtasks. Because of this possibility of divergence, goals may become individualized. Aggregate tasks call for a high degree of coordination among group members, even if members are doing essentially the same task within a project. For example, even if all members of the same group are conducting the same telephone survey, coordination is needed to ensure that the same questions are being asked, different respondents are being polled, and the statistics are similarly analyzed.

Projects assigned to collaborative writing classes are often aggregate tasks. Several years ago, my colleagues and I described an assignment we used in our business writing classes (Morgan, Allen, Moore, Atkinson, & Snow, 1987, pp. 20-26). This assignment required that a group of students plan and write four documents within a problem-solving writing unit. In fact, the task itself did not require a group to complete; any one student could—and did before we converted it to a group project—work alone. Within this project, any group of students could have assigned one document to each of the four members of the group, although this never happened to any of us. Or one student could have done all the work herself, although this never happened either. The assignment was typically a problem-solving one in that students were required to identify a problem in an organization and research solutions to the problem. Some aspects of discussion tasks were included in that students generated several problems through brainstorming and had to negotiate which problem the group would investigate. Agreement was often difficult to achieve.

Another aggregate assignment has been described by Blicq (1989). This assignment required three-member groups to choose a portable computer and make purchase recommendations to a department chair. The groups wrote five documents and made one oral presentation. As in the Morgan et al. assignment, nothing inherent in Blicq's assignment prevents his students from completing the task individually. His, however, is totally a problem-solving and not a discussion task: The issue is clear-cut and students have little need to reach consensus because the problem is presented by the instructor.

One of the most famous group assignments is the Tinkertoy assignment in which a group of students must assemble a mechanism from a box of Tinkertoys, then write a description of the mechanism and assembly instructions (Louth, 1989, pp. 229-234). In addition to time, the group advantage comes (1) when members pool their ideas and then agree on what to build and (2) when almost simultaneously they must manipulate the Tinkertoy pieces and write the description and directions. The invention aspect of the Tinkertoy project makes it a production task; the students are creating an image, albeit a three-dimensional image, that they must describe in words.

Interdependent Tasks

An interdependent task is a divisible task (as is the aggregate task), but it is conjunctive—all members must contribute to complete the assignment successfully. Allen et al.'s (1987) specialization group works on interdependent tasks: The group is assembled because of the special expertise of each member, and all members must contribute to the task or it will not be completed successfully. Interdependent tasks can have components that resemble either discussion tasks—such as generating and negotiating ideas—or problem-solving tasks. Cooperation requirements are even more stringent for interdependent tasks than for aggregate tasks because accomplishing the task depends on every member's input.

Many collaborative writing assignments in professional writing classes are interdependent tasks. Interdependence is created by assigning roles to each student in the group and successful completion depends on each student performing his or her role. Scott (1988) describes an assignment in which a group of students must invent something (pp. 138-142). She assigns roles to each student in the group—technical editor, writer, feasibility researcher, market planner, and communications specialist—with each student responsible for a critical portion of the whole project.

Raven (1990) describes a similar project in which a group must generate a new product or service and write a business plan persuading venture capitalists to invest in the idea. Although she does not assign roles to each group member, Raven does "require that each student take responsibility for a given section or sections of the plan" (p. 129).

Tackach (1991) teaches students how to manage a document using teams to guide "it through the entire publication process from planning and storyboarding, to writing, editing, and evaluating" (p. 118). In this assignment, which takes a class period to complete, specialized groups are formed that represent engineers, writers, illustrators, editors, and quality assurance personnel. The written product is a set of instructions for assembling and using a mechanical device. This assignment modifies the typical interdependent task in that each group performs only one function instead of group members performing separate functions. Groups, however, are working against the clock as Tackach gives the project only 90 minutes from initial design to final verification.

Guidelines for Collaborative Assignments

I would like to offer some brief guidelines for the design of collaborative writing assignments in professional writing classes based on the schema I described. First, if one of our purposes for using collaborative writing in our classrooms is to replicate experiences students will have in the workplace, then it is not in their best interest to assign individual tasks; few companies can afford to use groups on simple or routine tasks that easily can be completed by one person.

Second, the three other task types serve a variety of purposes for students and teachers. Melding and interdependent tasks make it impossible for one group member to do all the work while other members do little. Forcing whole-group participation by designing tasks in which all members must perform may provide a better educational experience than allowing students to choose whether or not they want to be involved, a problem many of us who use collaboration often face.

Third, we should use melding tasks more than we do. These tasks can be designed to address controversial issues or values relevant to professional writing. Putnam (1986) cites others to affirm the value of conflict; conflict "expands the range of judgments, engenders creative ideas, leads to a reexamination of opinions and goals, increases calculated risks, and fosters acceptance of the group decisions" (p. 177). Trimbur, speaking of collaborative writing and learning, urges teachers to encourage dissensus as a way of democratizing learning. Through what he calls a "rhetoric of dissensus," students can learn not only to accept but to value differences (Trimbur, 1989, p. 610). Topics such as racial quotas or sexual harassment, the use of nonsexist language, the impact of international politics on American corporations, or jobs versus anti-pollution measures—all appropriate in a professional writing class— can move students into areas vital to them, where the issues are not black and white, where there is no right or wrong, where *truth* is only probable. Assigning position papers, policies and procedures manuals, legal briefs, or recommendation reports based on such issues will encourage students to discuss their values, to recognize that other values exist, and to work through disagreements to consensus.

Conclusion

Professional writing teachers may use this schema to ensure that they offer a variety of collaborative writing experiences for their students, rather than designing one type of assignment that may not require students to use a full range of group experiences. Using tasks based on the schema can thus provide our students varied opportunities to create shared meanings through collaboration. The schema also provides a piece of a middle-range theory for teaching collaborative writing. It can provide a lens through which we evaluate both the assignments we create and those we read about in the literature on collaborative writing.

INTERPRETATION

14

Viewing Functional Pictures in Context

CHARLES KOSTELNICK

When readers construct meaning from professional documents, that activity encompasses not only the verbal text but *visual* elements as well. Like writing, visual communication also occurs in context, and thus it may also be regarded as a complex *social act*. Visual communication includes several areas of document design such as typography and text design, graphs and charts, and pictorial elements. Here I wish to focus solely on pictorial communication, which can be more clearly distinguished from the text than typography or graphics and can, therefore, help clarify how visual design embodies a social dimension.

A picture exists to communicate information; it is essentially, as Arnheim (1969) observes, "a statement about visual qualities, and such a statement can be complete at any level of abstractness" (p. 137). How does social context transform a pictorial *statement*? Although picture *recognition* is a fairly universal skill, *reading* pictures, like reading text, depends on contextual variables. In professional communication, functional pictures serve a variety of purposes—to describe objects and

AUTHOR'S NOTE: I wish to acknowledge my research assistant, Rick Dickson, for his help in gathering publications on visual communication.

mechanisms, to explain processes and methods, to persuade readers, to stimulate interest, to build trust—many of the same functions that the text performs in making information usable to readers. Like verbal communication, pictures represent an understanding of the world acquired by members of a certain group, and thus the meaning readers construct from a given image may depend largely on knowledge they share with group members.

Reading pictures extends far beyond a momentary, private response to a printed image, because readers filter pictorial information through a social lens. To explore the social nature of picture reading, I will briefly examine how picture perception is shaped by prior knowledge and experience. Then I will examine three levels of social context at which readers use their knowledge and experience to read functional pictures: (1) cultural context, which includes the shared worldview, experiences, or values of group members; (2) conventional context, in which readers share the visual language of a discipline or a specialized subject; and (3) immediate context, the constraints of the situation in which discourse participants use a picture.

Pictorial Perception and Knowledge

Picture reading depends on prior experience and knowledge, though the extent to which this is so has been the subject of inquiry and debate. Several researchers (Hagan & Jones, 1978, p. 172; Novitz, 1977, pp. 25-26; Perkins, 1980, pp. 261-262; see also Kennedy, 1974; Kostelnick, 1989) have defined this issue by citing the differing approaches of Gibson (1954) and Goodman (1976). Gibson held that picture perception is similar to the perception of real-world objects, while Goodman argued that picture perception is essentially a learned response. Both positions explain pictorial perception. Supporting Gibson's position, Jones and Hagan (1980) argue that "naive" readers of pictures—people who are unfamiliar with pictures and who typically live in remote cultures—need little experience to recognize pictures. Such is the case because pictures rendered in perspective enable readers to see objects and places similar to the way they view these things in the real world (pp. 194-195, 222).

While simple recognition of objects or places depicted pictorially may be a fairly universal skill, understanding the information in pictures is a more complicated business. Reading pictures involves not only what we see but what we know, and hence learning the visual language of pictures entails a process of initiation. Goodman (1976) rejects the notion of the "innocent eye" (pp. 7-8) and argues instead that pictures embody conventional languages that change from one culture or period to the next. Thus he considers even *realism* a convention, a pictorial style that readers learn to interpret (pp. 16, 34-39). Along similar lines, Gombrich (1972, p. 86) contends that pictures have a "code" that enables readers to grasp their meaning. A code suggests socially shared knowledge, and certainly great works of art embody complex codes. Even when we read functional pictures, however, we need to know the code to process the information. Although simple picture recognition may be a fairly universal skill, reading functional pictures is a profoundly learned response.

Experience strongly influences our perception of pictures because it is based on "knowledge" (Barnard & Marcel, 1984, p. 42; Gombrich, 1972, p. 86) built up through previous perceptual experiences. Palmer (1975, p. 280) shows how our knowledge of the world engenders "expectations" that help us structure visual information. While we can assume that readers who share similar perceptual experiences will share similar pictorial expectations, perceptual knowledge has more explicit social roots. It can be embedded in cultural values and knowledge, in the conventional codes shared by members of a profession or discipline, or through the interaction among senders and receivers in a particular situation. This pictorial knowledge is profoundly social because it is shared among members of a group: members of cultures, members of disciplines or professions, and immediate discourse participants. Below, I explore how these three social contexts affect our reading of functional pictures.

Cultural Context: The Widest Lens

Cultural knowledge creates the widest social lens through which we read pictures. Variations in picture perception across cultures have been widely studied and analyzed (see Deregowski, 1980, Hagan & Jones, 1978; Jones & Hagan, 1980). Readers' interpretations of pictures, however, entail cultural factors that extend beyond the perceptual mechanics of eye and image. As Fresnault-Deruelle (1983, p. 13) points out, "contexte" is "visible" as well as "non-visible", the latter deriving from

Figure 14.1. Illustration From *The Mirrour of the World*
SOURCE: Caxton (1490) Courtesy of the Rare Book and Special Collections Library at the University of Illinois at Champaign-Urbana.

historical and cultural conditions in which an image is situated. This unseen *contexte* is essentially social because it is based on the collective values and beliefs of viewers.

Certainly culture and worldview influence our reading of the picture in Figure 14.1, from Caxton's (1490) translation and printing of *The Mirrour of the World,* one of the earliest printed texts explaining scientific theories (see Prior, 1913, pp. v-vi). Here we have four figures preparing to drop stones into the earth, all of which will presumably meet at the center. On a level of literal recognition, we can probably identify key images in the picture—even though the figures are somewhat medieval in their proportions and clothing, the holes in the earth are ridiculously large, and three of the four figures appear to be falling off the planet. Still, even though the picture lacks the naturalistic perspective of Renaissance drawing, we can recognize key elements. For modern readers, however, *identification* falls far short of *comprehension* because the worldview that underpins this picture, in which everything in the universe points inwardly toward the earth, obscures its meaning. The picture is a cultural and historical artifact because it is based on an outmoded, anthropocentric worldview that locates the earth at the center of the universe, circled by the elements, the ether world, the planets, and the sun. Needless to say, the logic of this picture eludes us today because its cultural and scientific bases no longer hold.

In Caxton's picture the image and the worldview underpinning the image seem inseparable. A more "realistic" picture, however, cannot guarantee clarity of meaning, even though the pictorial style may appear to be more accessible to modern readers. Figure 14.2, from Langley's (1728) *New Principles of Gardening,* is essentially an instructional text, providing guidelines and examples for laying out gardens. Forms in the picture are arranged in symmetrical alleys and geometrical configurations, revealing a classical attitude toward nature. However, a tension exists between these controlled, classical forms and the irregular (and irrational) forms of the picturesque, embodied in the artificial ruin at the focal point of the perspective. Artificial ruins were placed in gardens to create an affecting picture by evoking the mood of the past. Here we have a picture within a picture, and one that is pro- foundly social because its messages are intertwined with the cultural setting that produced it.

Pictures are imbued with a certain cultural knowledge that reflects the shared experience of viewers at a certain historical moment. Because this experience varies among groups of viewers, and over time, conveying practical information pictorially for large, multicultural audiences presents formidable barriers. Communicating statistical data, creating symbols for corporations, designing informational signs for airports, highways, or other public places—these design tasks require a great deal of sensitivity to cultural context. One of the most ambitious attempts to develop a universal picture language was undertaken by Neurath (1936/1980) in the 1930s. Neurath wanted his International System of Typographic Picture Education (ISOTYPE) to break the barriers of verbal language by providing a precise and objective mode of communication. Today, however, these timeless characteristics give way to the cultural associations of the period, partly because of ISOTYPE's

Figure 14.2. Garden Scene From *New Principles of Gardening*

SOURCE: Langley (1728, Plate XXII). Courtesy of Iowa State University Library—Department of Special Collections.

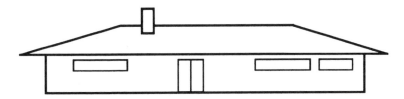

Figure 14.3. Ranch-Style House

affinities with early 20th-century modernism (see Lupton, 1986, p. 50), especially the Bauhaus and the International Styles.

The difficulty of designing a universal picture language shows the importance of cultural knowledge in pictorial perception. This knowledge, which gives readers access to the science in Caxton's picture or the gardening methods in Langley's drawing, is the widest social lens through which pictorial perception occurs. Hence pictures that might be appropriate for one setting might look exotic or dated in another or might offend or confuse viewers who see them from a different cultural perspective. Cultural context filters pictorial information even when we view the most mundane image. If I placed the drawing in Figure 14.3 in an instructional brochure for homeowners, Midwestern readers would probably associate this form with a typical dwelling, because many ranch-style houses have been built in the Midwest since World War II. People in other regions of the world (or even of the United States) might not have the same affinity with this form, and thus they might not understand or value its information because they do not dwell—nor might they wish or be able to dwell—in this style of house. For these readers, the picture might diminish the credibility of the brochure or simply be regarded as anomalous.

Conventional Context:
Discipline-Specific Codes

While cultural knowledge provides the largest umbrella for social context, communities of picture readers use another kind of shared knowledge—pictorial conventions. Conventions for rendering pictures can be broadly defined by culture (see Mangan, 1978), but more important for functional pictures they can be shared by members of certain professions or disciplines (see Ashwin, 1984, p. 51). Pictorial conventions, especially in technical pictures, situate readers in a world that they know, or that they want to know. Even readers who do not share the same cultural knowledge can share a common understanding of discipline-specific conventions. Visual codes used in electrical diagrams, architectural plans, mechanical drawings, and the like can easily bridge cultural boundaries by establishing a kind of visual discourse community. Because these conventional languages are agreed on among users, and sometimes sanctioned and reviewed by professional organizations, they also embody an important social function.

Discipline-specific conventions define audiences by performing a gatekeeping function that allows some readers access to information while restricting other readers. Conventions, however technical, require readers to tap into their previous experience with pictorial information: Readers process information accessible to them, make some guesses about information they cannot process, and ignore (or misinterpret) unconventional information. Figure 14.4, from *Appleton's Cyclopaedia of Drawing* (Worthen, 1869), is a fairly technical picture of a stop gate mechanism that regulates the flow of water. Here the visual language—the graphic coding, the level of detail, the spatial arrangement—limit and define the audience. The graphic coding (the cross-sectional lines, the tiny circles, the dotted lines) might perplex some readers or simply represent different things in different disciplines. The level of detail also defines the audience because it provides clues about the kinds of information readers understand and value. The details included in the picture—bolts, threads on a shaft, specific dimensions—might be useful to an engineer, a manufacturer, or a maintenance superintendent but be meaningless to anyone else. Spatially, the drawing cuts across two different planes: The left half of the drawing is rendered in cross-section; the right half, in elevation (Worthen, 1869, p. 203), a picturing technique that reveals additional information but that further removes this image from the nontechnical reader.

Using pictorial conventions can be particularly problematic in instructional materials intended for lay audiences. While technical readers may be used to seeing, or thinking about, objects displayed at certain angles that reveal different bits of information, nontechnical readers may not be accustomed to using pictures this way. Barnard and Marcel (1984) discuss an example from a washing machine manual that showed a detail of the appliance from the top rather than the side, a spatial orientation that could confuse readers (pp. 67-68). Moreover,

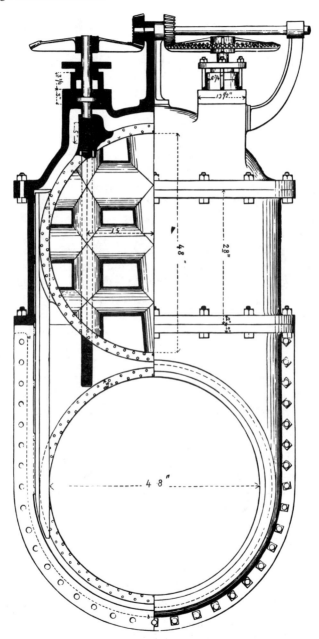

Figure 14.4. Elevation/Section From *Appletons' Cyclopaedia of Drawing*
SOURCE: Worthen (1869, p. 202; Plate XXXVIII).

the convention of rendering an object in a flat plane without any per-spective—e.g., in a *plan, elevation,* or *cross-section*—is foreign to most laypersons who do not routinely use these kinds of pictures and who certainly do not see reality that way. When an architect or engineer re-views a set of plans with a client, the professional and the client read images differently because the client has far less experience translating flat planes into real space. That is not to say that functional pictures always have to match the reader's perception of the real world: The users of the stop gate drawing in Figure 14.4 will probably never see this object in cross-section, but because of their technical knowledge they can still apply the information in the picture.

Pictures contain stylistic features that define the purpose of the com-munication, who the readers are, and how readers ought to use the in-formation. These stylistic qualities include level of technicality, the level of abstraction, and tone—all concepts from writing that also apply to pictures. The picture in Figure 14.5, from a document issued by the Iowa State University Extension Service instructing readers on how to build a compost pile, is far less formal than Figure 14.4. The informal tone and low level of technicality show that the picture was intended for a wide audience who ultimately will *do* something with the infor-mation. Despite stylistic features that widen the audience, the picture reveals some assumptions about the kinds of knowledge readers share. Readers must grasp the spatial and graphic conventions of a cross-section: just as in the stop gate drawing, half of the picture is shown in cross-section and half in elevation. This spatial dichotomy, however, not only supplies additional information but also helps readers identify the convention of the cross-section, which they might not be accus-tomed to seeing. Readers of this picture must also understand the con-ventional functions of the arrows, the one on the right signaling mea-surement, the others connecting explanatory text to features of the pic-ture. These visual conventions may not impede most readers, but they require readers to tap into their prior picture reading experience to trans-form this image into useful information.

Immediate Context:
Pictures in Situations

Conventions socialize pictorial communication because they define groups that understand codes, spatial arrangements, and details that

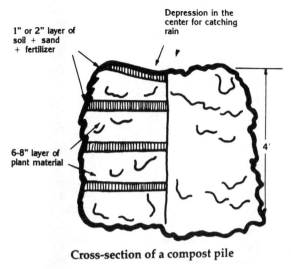

Cross-section of a compost pile

Figure 14.5. Illustration From *Composting Yard Waste*
SOURCE: Taber, Naeve, Agnew, and Heer (1991). Reprinted by permission of Iowa State University
Extension Communication Services.

those outside the group do not comprehend, or at least comprehend as
fully, as group members. Pictures have a yet more immediate social
dimension: as a transaction between the sender and receiver in a given
situation for a particular purpose. Although we sometimes draw pictures
on the spot (e.g., a map for someone who is lost), pictures are not as
spontaneous as spoken language because, like text, they take time to
create. However, as a "statement about visual qualities," as Arnheim
(1969) puts it, a picture embodies a rich, flexible language. If we consider the use of pictures, as Novitz (1977) does, we can see more clearly
how pictures fulfill the purposes and expectations of a particular communicative act.

The purposes of pictorial language have been analyzed by Ashwin
(1984), Goldsmith (1984), Novitz (1977), and Twyman (1985). Taking
a semiotic approach, Ashwin has shown how pictures serve a variety of
purposes—referential, poetic, conative, emotive, etc. Drawings of a new
piece of equipment that engineers study for hours at a drafting table
would provide highly referential information, with the images specify-
ing the precise location of each component. Like written information,
pictures have a variety of purposes depending on the subject, reader,
and situation. The stop gate drawing in Figure 14.4, for example,

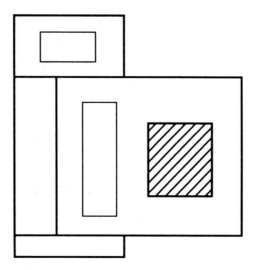

Figure 14.6. Sample Illustration

obviously serves a referential purpose; however, it could also serve
emotive or poetic purposes if I used it to demonstrate compositional
balance to design students (see Ashwin, 1984, pp. 47, 50-51). Pictures
are extremely elastic because, depending on the situation, the same picture can serve a variety of purposes (Novitz, 1977, pp. 5-10). For example,
I could use Figure 14.1 on a brochure to stimulate interest in a literary
or scientific conference, Figure 14.2 to illustrate visual contrast or irony
(an irregular ruin in a rectilinear garden), or Figure 14.3 to market home
insurance or a new line of windows.

These purposes are often determined by the situational contexts in
which pictures are actually used. Depending on the situation, the image
I created for Figure 14.6 could serve as a logo for an organization, a plan
for a new office complex, or a proposal for a new tool. If it were the
latter, I would need to know what readers are going to do with the tool:
manufacturers who might make it, investors who might finance its development, technicians who might repair it, consumers who might buy
it, and so on. Like language users, users of pictures achieve their purposes by fulfilling the needs and expectations of their intended readers:
Researchers who will use the tool in a laboratory will look for precise
specifications and measurements, while managers interested in marketing the tool to homeowners will focus on surface features, design

appeal, and the like. Adapting the picture to meet these expectations will help ensure a successful transaction between a sender and receiver.

Tailoring pictures to specific uses creates a social context because the designer acquiesces to the needs and expectations of readers. Just as immediate context shapes spoken or written discourse, pictures are immersed in particular settings and reshaped by the social dynamics of these settings. As Novitz (1977, pp. 67-85) demonstrates, pictures sometimes function like *illocutionary acts* because their pragmatic use is defined by the interaction between sender and receiver in a given situation (see also Barnard & Marcel, 1984, pp. 40-41, 60). Depending on the situation, the statement "I want a copy of your report" could be construed as a routine request (for a copy of a subordinate's document), a threat (challenging the writer's conclusions), or a compliment (a speaker wanting to emulate the writer's document). Meaning here depends on the social interaction between speaker and listener—the previous communication between the two, the authority of one over another, the organizational factors within a department or group, and so on.

These kinds of situational factors also transform pictorial perception. If an office manager used Figure 14.6 in an instructional memo to nontechnical employees with low morale, its sparse details, lack of perspective, and spatial orientation (top view) might confuse employees and, in this situation, engender resentment and even cynicism. More details, less technical coding, and a more casual tone (perhaps by drawing the image freehand) might better suit this context, and importantly, reassure readers that the manager recognizes their concerns. On the other hand, if the office manager used the same drawing in a memo to senior engineers who suspect the manager's technical expertise, the abstractness and low level of technicality of the drawing might further erode their confidence. The manager could use details like those in Figure 14.4 to shore up their trust, though another group of readers—say newly hired staff engineers, less secure in their jobs and out to prove themselves—might view the picture as an encroachment on their area of expertise. Only by knowing the immediate situation can users anticipate these responses and adapt pictures to them.

Many other scenarios, of course, are possible. Perhaps a land developer wants to use the picture to show investors the site plan for a new shopping mall. In this case, supplying only sparse details might be the best strategy, creating a mutual understanding with readers that the picture represents only a fraction of the information already displayed in the plans. In another situation such scant information might make readers nervous or suspicious; however, in this case the picture makes a diplomatic bow to the investors by acknowledging that they are too important or busy to be bogged down with details, that they make only high-level decisions, not trivial ones. Like language, pictures are mediated by the social dynamics of the situations in which senders and receivers find themselves. Tactful users of pictures need to consider these circumstances, just as writers do, to meet their objectives.

Conclusion

Reading functional pictures is a social rather than an isolated individual response to visual images. This social dimension provides a common bond between pictorial and verbal communication and has clear implications for practice: Professional communicators must adapt pictures, just as they do text, to the experience and expectations of readers. Guided by the reader's knowledge, perception begins before the visual encounter with a given image. This knowledge can be shared on many levels—broadly across a culture, among members of a discipline, or locally by immediate discourse participants. All of these contextual levels mediate perception by broadening the foundation for picture reading beyond the experience of the individual viewer. The process of reading pictures can be understood only with reference to these contextual variables because, like language, pictures do not communicate information in a vacuum.

References

Acker, J. (1991). Hierarchies, jobs, bodies: A theory of gendered organizations. In J. Lorber & S. A. Farrell (Eds.), *The social construction of gender* (pp. 162-179). Newbury Park: Sage.

Aires, E. (1987). Gender and communication. In P. Shaver & C. Hendrick (Eds.), *Sex and gender* (pp. 149-176). Newbury Park, CA: Sage.

Allen, N., Atkinson, D., Morgan, M., Moore, T., & Snow, C. (1987). What experienced collaborators say about collaborative writing. *Journal of Business and Technical Communication, 1*(1), 70-90.

Alpers, S. (1983). *The art of describing: Dutch art in the seventeenth century.* Chicago: University of Chicago Press.

Amsterdamska, O., & Leydesdorff, L. (1989). Citations: Indicators of significance? *Scientometrics, 15,* 449-471.

Anderson, P. V. (1985). Survey methodology. In L. Odell & D. Goswami (Eds.), *Writing in nonacademic settings* (pp. 453-501). New York: Guilford.

Anderson, P. V., Brockmann, J., & Miller, C. R. (Eds.). (1983). *New essays in technical and scientific communication: Research, theory, and practice.* Farmingdale, NY: Baywood.

Anson, C. M., & Forsberg, L. L. (1990). Moving beyond the academic community. *Written Communication, 7*(2), 200-231.

Althusser, L. (1971). Ideology and the State. In *Lenin and philosophy and other essays* (B. Brewster, Trans., pp. 123-173). London: New Left Books.

Aristotle. (1976). *The ethics of Aristotle: Nicomachean ethics* (J. A. K. Thomson, Trans.). London: Penguin.

Aristotle. (1984). *Rhetoric* (F. Solmsen, Ed.; W. R. Roberts, Trans.). In B. Jowett (Ed.), *The rhetoric and poetics of Aristotle* (pp. 3-218). New York: Modern Library. (Original work c.323 B.C.)

Arms, V. M. (1987). Engineers becoming writers: Computers and creativity in technical writing classes. In L. Gerrard (Ed.), *Writing as century's end: Essays on computer-assisted composition* (pp. 64-78). New York: Random House.

Arnheim, R. (1969). *Visual thinking.* Berkeley: University of California Press.

257

Arnheim, R. (1974). *Art and visual perception: A psychology of the creative eye.* Berkeley: University of California Press.

Ashwin, C. (1984). Drawing, design, and semiotics. *Design Issues, 1*(2), 42-52.

Atkins, D. G. (1990, Fall). The return of/to the essay. *Association of Departments of English Bulletin, 96,* 11-18.

Bachrach, S., & Baratz, M. (1962). Two faces of power. *American Political Science Review, 56,* 947-952.

Bacon, F. (1620/1955). Novum organum. In H. G. Dick (Ed.), *The works of Francis Bacon* (pp. 455-540). New York: Modern Library.

Baird, J. E., Jr. (1976). Sex differences in group communication: A review of relevant research. *Quarterly Journal of Speech, 62*(1), 179-192.

Baird, J. E., Jr., & Bradley, P. H. (1979). Styles of management and communication: A comparative study of men and women. *Communication Monographs, 46*(2), 101-111.

Bakhtin, M. M. (1981). *The dialogic imagination: Four essays* (M. Holquist, Ed.; C. Emerson & M. Holquist, Trans.). Austin: University of Texas Press.

Bakhtin, M. M. (1986). The problems of speech genres. In C. Emerson, M. Holquist, & V. Mcgee (Eds. and Trans.), *Speech genres and other late essays* (pp. 60-102). Austin: University of Texas Press.

Bales, R. F., & Strodtbeck, F. L. (1951). Phases in group problem solving. *Journal of Abnormal and Social Psychology, 46,* 485-495.

Barnard, P., & Marcel, T. (1984). Representation and understanding in the use of symbols and pictograms. In R. Easterby & H. Zwaga (Eds.), *Information design: The design and evaluation of signs and printed materials* (pp. 37-75). New York: Wiley.

Barnes, L. L. (1990). Gender bias in teachers' written comments. In S. L. Gabriel & I. Smithson (Eds.), *Gender in the classroom: Power and pedagogy* (pp. 140-159). Urbana: University of Illinois Press.

Barthes, R. (1970). *Mythologies* (A. Lavers, Trans.). New York: Hill & Wang.

Barthes, R. (1985). *The grain of the voice: Interviews 1962-1980* (L. Coverdale, Trans.). New York: Hill & Wang.

Barthes, R. (1986). *The rustle of language* (R. Howard, Trans.). New York: Hill & Wang.

Bartholomae, D. (1985). Inventing the university. In M. Rose (Ed.), *When a writer can't write* (pp. 134-165). New York: Guilford Press.

Barton, B. F., & Barton, M. S. (1987). Simplicity in visual representation: A semiotic approach. *Journal of Business and Technical Communication, 1*(1), 9-26.

Barton, B. F., & Barton, M. S. (1989). Trends in visual representation. In C. H. Sides (Ed.), *Technical and business communication: Bibliographic essays for teachers and corporate trainers* (pp. 95-135). Urbana, IL: National Council of Teachers of English and Society for Technical Communication.

Barton, B. F., & Barton, M. S. (1990). Postmodernism and the relation of word and image in professional discourse. *Technical Writing Teacher 17*(3), 256-270.

Baudrillard, J. (1983). *Simulations* (P. Foss, P. Patton, & P. Beitchman, Trans.). New York: Semiotext(e).

Bazerman, C. (1985). Physicists reading physics. Schema-laden purposes and purpose-laden schema. *Written Communication, 2*(1), 3-23.

Bazerman, C. (1988). *Shaping written knowledge: The genre and activity of the experimental article in science.* Madison: University of Wisconsin Press.

Bazerman, C. (in press). Linguistic and rhetorical studies of the disciplines. In T. Becher (Ed.), *The encyclopedia of higher education.* London: Pergamon Press.

Beard, J. D. (1990). A collaborative simulation. *Bulletin of the Association for Business Communication, 53*(2), 65-67.

Beard, J. D., & Rymer, J. (Eds.). (1990). Collaborative writing in business communication [Special issue]. *Bulletin of the Association for Business Communication, 53*(2).

Beard, J. D., Rymer, J., & Williams, D. L. (1989). An assessment system for collaborative-writing groups: Theory and empirical evaluation. *Journal of Business and Technical Communication, 3*(2), 29-51.

Belenky, M. F., Clinchy, B. M., Goldberger, N. R., & Tarule, J. M. (1986). *Women's ways of knowing: The development of self, voice, and mind.* New York: Basic Books.

Belsey, C. (1980). *Critical practice.* London: Methuen.

Bem, S. L. (1981). Gender schema theory: A cognitive account of sex typing. *Psychological Review, 88*(4), 354-364.

Berlin, J. (1988). Rhetoric and ideology in the writing class. *College English, 50,* 477-494.

Bertin, J. (1983). *Semiology of graphics: Diagrams, networks, maps* (W. J. Berg & P. Scott, Trans.). Madison: University of Wisconsin Press.

Beutler, B., Milsmark, I. W., & Cerami, A. C. (1985). Passive immunization against cachectin/tumor necrosis factor protects mice from lethal effect of endotoxin. *Science, 229,* 869-871.

Bezucha, R. J. (1985). Feminist pedagogy as a subversive activity. In M. Culley & C. Portuges (Eds.), *Gendered subjects: The dynamics of feminist teaching* (pp. 81-95). Boston: Routledge & Kegan Paul.

Bitzer, L. F. (1968). The rhetorical situation. *Philosophy and Rhetoric, 1*(1), 1-14.

Bizzell, P. (1986a). Composing processes: An overview. In A. R. Petrosky & D. Bartholomae (Eds.), *The national society for the study of education yearbook: The teaching of writing* (pp. 49-70). Chicago: University of Chicago Press.

Bizzell, P. (1986b). Foundationalism and anti-foundationalism in composition studies. *PRE/TEXT, 7,* 37-56.

Bizzell, P. (1986c, March). *Academic discourse: Taxonomy of conventions or collaborative practice?* Paper presented at the Conference on College Composition and Communication, New Orleans.

Bizzell, P. (1990). Beyond anti-foundationalism to rhetorical authority: Problems defining "cultural literacy." *College English, 52,* 661-675.

Bleich, D. (1988). *The double perspective: Language, literacy, and social relations.* New York: Oxford University Press.

Bleich, D. (1989). Genders of writing. *Journal of Advanced Composition, 9*(1), 10-25.

Bleich, D. (1990). Sexism in academic styles of learning. *Journal of Advanced Composition, 10*(2), 231-247.

Blicq, R. S. (1989). Evaluation of portable computers for the department. In R. Louth & A. M. Scott (Eds.), *Collaborative technical writing: Theory and practice* (pp. 189-192). Minneapolis: Association for Teachers of Technical Writing.

Booth, W. C. 1961. *The rhetoric of fiction.* Chicago: University of Chicago Press.

Borgès, J. L. (1972). *A universal history of infamy* (N. T. di Giovanni, Trans.). New York: Dutton.

Bosley, D. S. (1989). *A national study of the uses of collaborative writing in business communication courses among members of ABC.* Unpublished doctoral dissertation, Illinois State University.

Bosley, D. S. (1990). Individual evaluations in a collaborative report project. *Technical Communication, 37,* 160-162.

Bourdieu, P. (1977). *Outline of a theory of practice* (R. Nice, Trans.). Cambridge, UK: Cambridge University Press.

Brodkey, L. (1987). Writing ethnographic narratives. *Written Communication, 4,* 25-50.

Brooke, R. E. (1991). *Writing and sense of self: Identity negotiation in writing workshops.* Urbana, IL: National Council of Teachers of English.

Brooks, C., & Warren, R. P. (1970). *Modern rhetoric* (3rd ed.). New York: Harcourt Brace Jovanovich

Brown, R. L., Jr., & Herndl, C. G. (1986). An ethnographic study of corporate writing: Job status as reflected in written text. In B. Couture (Ed.), *Functional approaches to writing: Research perspectives* (pp. 11-28). Norwood, NJ: Ablex.

Bruffee, K. A. (1973). Collaborative learning: Some practical models. *College English, 34,* 634-643.

Bruffee, K. A. (1982). Liberal education and the social justification of belief. *Liberal Education, 68,* 95-114.

Bruffee, K. A. (1983). Writing and reading as collaborative social acts. In J. Hays, P. A. Roth, J. R. Ramsey, & R. D. Foulke (Eds.), *The writer's mind: Writing as a mode of thinking* (pp. 159-169). Urbana, IL: National Council of Teachers of English.

Bruffee, K. A. (1984). Collaborative learning and the "conversation of mankind." *College English, 46,* 635-652.

Bruffee, K. A. (1985). *A short course in writing: Practical rhetoric for teaching composition through collaborative learning* (3rd ed.). Boston: Little, Brown.

Bruffee, K. A. (1986). Social construction, language, and the authority of knowledge: A bibliographical essay. *College English, 48,* 773-790.

Bruner, J. (1978). The role of dialogue in language acquisition. In A. Sinclair, R. J. Jarvella, & W. J. M. Levelt (Eds.), *The child's conception of language* (pp. 241-256). New York: Springer-Verlag.

Bruner, J. (1991). The narrative construction of reality. *Critical Inquiry, 18,* 1-21.

Bryson, N. (1983). *Vison and painting.* New Haven, CT: Yale University Press.

Bryson, N. (1989). Review of *The interpretation of pictures,* by Mark Roskill. *The Art Bulletin, 71,* 704-707.

Burke, K. (1961). *The philosophy of literary form.* New York: Random House.

Burke, M., & McLaren, I. (1981). London's public transport diagrams—visual comparisons of some graphic conventions. *Information Design Journal, 2*(2), 103-112.

Burnett, R. (1990, June). Benefits of collaborative planning in the business communication classroom. *Bulletin of the Association for Business Communication, 53*(2), 9-17.

Burnett, R. (1991). Substantive conflict in a cooperative context: A way to improve the collaborative planning of workplace documents. *Technical Communication, 38,* 532-539.

Campbell, G. (1776/1963). *The philosophy of rhetoric* (L. F. Bitzer, Ed.). Carbondale: Southern Illinois University Press. (Original work published 1776)

Carter, M. (1990). The idea of expertise: An exploration of cognitive and social dimensions of writing. *College Composition and Communication, 41,* 265-286.

Caxton, W. (Ed. and Trans.). (1490). *The mirrour of the World.* Westminster: Author.

Chase, G. (1988). Accomodation, resistance and the politics of student writing. *College Composition and Communication, 39,* 13-22.

Chodorow, N. (1978). *The reproduction of mothering: Psychoanalysis and the sociology of gender.* Berkeley: University of California Press.

Christian, S. E. (1990, November 14). Atlas cartographers keep their erasers handy. *The Ann Arbor News,* p. C1.

Christie, F. (1985). Language and schooling. In S. Tchudi (Ed.), *Language, schooling, and society* (pp. 21-40). Upper Montclair, NJ: Boynton.

Christie, F. (1987). Genre as choice. In I. Reid (Ed.), *The place of genre in learning* (pp. 22-34, Typereader Publications no. 1). Geelong, Australia: Deakin University Press, The Centre for Studies in Literary Education.

Clark, G. (1990). *Dialogue, dialectic, and conversation: A social perspective on the function of writing.* Carbondale: Southern Illinois University Press.

Clark, G. (1987). Ethics in technical communication: A rhetorical perspective. *IEEE Transactions on Professional Communication, 30*(3), 190-195.

Clark, H. H., & Haviland, S. H. (1977). Comprehension and the Given-New Contract. In R. O. Freedle (Ed.), *Discourse production and comprehension* (pp. 1-40). Norwood, NJ: Ablex.

Clark, S., & Ede, L. (1990). Collaboration, resistance, and the teaching of writing. In A. Lunsford, H. Moglen, & J. Slevin (Eds.), *The right to literacy* (pp. 276-285). New York: Modern Language Association.

Clifford, J. (1983). On ethnographic authority. *Representations, 1*(2), 118-146.

Colvin, M. (1990, September 10). Letter from Baghdad: A war-weary people watch—and hope. *U.S. News & World Report,* p. 32.

Comprone, J. (1990). The literacies of science and humanities: The monologic and dialogic traditions. In K. Ronald & H. Roskelly (Eds.), *Farther along: Transforming dichotomies in rhetoric and composition* (pp. 52-70). Portsmouth, NH: Boynton.

Cooke, P. (1990). *Back to the future: Modernity, postmodernity, and locality.* London: Unwin Hyman.

Cooper, M. M. (1986). The ecology of writing. *College English, 48,* 364-375.

Cooper, M. M. (1989). Why are we talking about discourse communities? In M. M. Cooper & M. Holzman, *Writing as social action* (pp. 202-220). Portsmouth, NH: Boynton.

Cooper, M. M., & Holzman, M. (1989). *Writing as social action.* Portsmouth, NH: Boynton.

Couture, B. (Ed.). (1986). *Functional approaches to writing: Research perspectives.* London: Frances Pinter.

Couture, B., & [Rymer] Goldstein, J. (1984). Procedures for developing a technical communication case. In R. J. Brockmann et al. (Eds.), *The case method in technical communication: Theory and models* (pp. 33-46, Association of Teachers of Technical Writing Anthology Series). Minneapolis: Association of Teachers of Technical Writing.

Couture, B., & [Rymer] Goldstein, J. (1985). *Cases for technical and professional writing.* Boston: Little, Brown.

Couture, B., & Rymer, J. (1989). Interactive writing on the job. In M. Kogen (Ed.), *Writing in the business professions* (pp. 73-93). Urbana, IL: National Council of Teachers of English, Association for Business Communication.

Couture, B., & Rymer, J. (1991). Discourse interaction between writer and supervisor. In M. M. Lay & W. M. Karis (Eds.), *Collaborative writing in industry* (pp. 87-108). Amityville, NY: Baywood.

Cozzens, S. E. (1985). Comparing the sciences: Citation content analysis of papers from neuropharmacology and sociology of science. *Social Studies of Science, 15,* 127-153.

Cozzens, S. E. (1989). What do citations count? The rhetoric-first model. *Scientometrics, 15,* 437-447.

Dasenbrock, R. W. (1991). Do we write the text we read? *College English, 53*(1), 7-18.

Daumer, E., & Runzo, S. (1987). Transforming the composition classroom. In C. L. Caywood & G. Overing (Eds.), *Teaching writing: Pedagogy, gender, and equality* (pp. 45-62). Albany: State University of New York Press.

Davidson, D. (1986a). On the very idea of a conceptual scheme. In *Inquiries into truth and interpretation* (pp. 183-198). Oxford, UK: Clarendon.

Davidson, D. (1986b). Communication and convention. In *Inquiries into truth and interpretation* (pp. 265-280). Oxford, UK: Clarendon.

Davidson, D. (1986c). A coherence theory of truth and knowledge. In E. LePore (Ed.), *Truth and interpretation: Perspectives on the philosophy of Donald Davidson* (pp. 307-319). New York: Blackwell.

Davidson, D. (1986d). A nice derangement of epitaphs. In E. LePore (Ed.), *Truth and interpretation: Perspectives on the philosophy of Donald Davidson* (pp. 433-446). New York: Blackwell.

de Certeau, M. (1984). *The practice of everyday life* (S. Rendall, Trans.). Berkeley: University of California Press.

de Saussure, F. (1966). *Course in general linguistics* (Eds. C. Bally & A. Sechehaye, trans. W. Baskin). New York: McGraw-Hill.

Deal, T., & Kennedy, A. A. (1982). *Corporate cultures: The rites and rituals of corporate life.* Reading, MA: Addison-Wesley.

Dell, S. A. (1990). Promoting equality of the sexes through technical writing. *Technical Communication, 37,* 248-251.

DeMott, B. (1990). *The imperial middle: Why Americans can't think straight about class.* New York: William Morrow.

Deregowski, J. B. (1980). *Illusions, patterns and pictures: A cross-cultural perspective.* London: Academic Press.

Derrida, J. (1976). *Of grammatology* (Trans. G. C. Spivak). Maryland: Johns Hopkins University Press.

Derrida, J. (1979). Living on: Borderlines. In H. Bloom (Ed.), *Deconstruction and criticism* (pp. 75-176). New York: Seabury.

Dieter, O. A. L. (1950). Stasis. *Speech Monographs, 17,* 345-369.

DiPardo, A., & Freedman, S. W. (1988). Peer response groups in the writing classroom: Theoretic foundations and new directions. *Review of Educational Research, 58,* 119-149.

Dissoi Logoi. (1972). In R. K. Sprague (Ed. and Trans.), *The older Sophists.* Columbia: University of South Carolina Press. (Original work c.420 B.C.)

Dixon, J. (1987). The question of genres. In I. Reid (Ed.), *The place of genre in learning* (pp. 9-21, Typereader Publications no. 1). Geelong, Australia: Deakin University Press, The Centre for Studies in Literary Education.

Doheny-Farina, S. (1986). Writing in an emerging organization: An ethnographic study. *Written Communication, 3,* 158-185.

Doheny-Farina, S. (Ed.). (1987). Legal and ethical aspects of technical communication: A special issue [Special issue]. *IEEE Transactions on Professional Communication, 30*(3).

Doheny-Farina, S. (1989a). Ethics and technical communication. In C. Sides (Ed.), *Technical and business communication: Bibliographic essays for teachers and corporate trainers* (pp. 53-73). Urbana, IL: National Council of Teachers of English and Society for Technical Communication.

Doheny-Farina, S. (1989b). A case study of one adult writing in academic and nonacademic discourse communities. In C. M. Matalene (Ed.), *Worlds of writing: Teaching and learning in discourse communities of work* (pp. 17-42). New York: Random House.

Driskill, L. (1989). Understanding the writing context in organizations. In M. Kogen (Ed.), *Writing in the business professions* (pp. 125-145). Urbana, IL: National Council of Teachers of English and Association for Business Communication.

Duin, A. H. (1991). Computer-supported collaborative writing: The workplace and the writing classroom. *Journal of Business and Technical Communication, 5,* 123-150.

Duin, A. H., Jorn, L. A., & DeBower, M. S. (1991). Collaborative writing—Courseware and telecommunications. In M. M. Lay & W. M. Karis (Eds.), *Collaborative writing in industry: Investigations in theory and practice* (pp. 146-169). Amityville, NY: Baywood.

Ede, L. (1984). Audience: An introduction to research. *College Composition and Communication, 35,* 140-154.

Ede, L., & Lunsford, A. (1984). Audience addressed/audience invoked: The role of audience in composition theory and pedagogy. *College Composition and Communication, 35,* 155-171.

Ede, L., & Lunsford, A. (1990). *Singular texts/plural authors: Perspectives on collaborative writing.* Carbondale: Southern Illinois University Press.

Edgerton, S. Y., Jr. (1987). From mental matrix to *mappamundi* in Christian empire: The heritage of Ptolemaic cartography in the Renaissance. In D. Woodward (Ed.), *Art and cartography: Six historical essays* (pp. 10-50). Chicago: University of Chicago Press.

Eiler, M. A. (1989). Process and genre. In C. B. Matalene (Ed.), *Worlds of writing: Teaching and learning in discourse communities of work* (pp. 17-42). New York: Random House.

Emmerson, D. K. (1984). "Southeast Asia": What's in a name. *Journal of Southeast Asia Studies, 15,* 1-21.

Ewald, H. R., & MacCallum, V. (1990). Promoting creative tension within collaborative writing groups. *Bulletin of the Association for Business Communication, 53*(2), 23-26.

Faigley, L. (1985). Nonacademic writing: The social perspective. In L. Odell & D. Goswami (Eds.), *Writing in nonacademic settings* (pp. 231-248). New York: Guilford Press.

Faigley, L. (1986). Competing theories of process: A critique and a proposal. *College English, 48,* 527-542.

Farkas, D. (1991). Collaborative writing, software development, and the universe of collaborative activity. In M. M. Lay & W. M. Karis (Eds.), *Collaborative writing in industry: Investigations in theory and research* (pp. 13-30). Amityville, NY: Baywood.

Fearing, B. E., & Sparrow, W. K. (Eds.). (1989). *Technical writing: Theory and practice.* New York: Modern Language Association.

Fish, S. E. (1980). *Is there a text in this class? The authority of interpretive communities.* Cambridge, MA: Harvard University Press.

Fisher, W. R. (1987). *Human communication as narration: Toward a philosophy of reason, value, and action.* Columbia: University of South Carolina Press.

Fitzgerald, K. (1990, December). Whistle-blowing: Not always a losing game. *IEEE Spectrum, 27,* 49-52.

Flower, L., & Hayes, J. R. (1981). A cognitive process theory of writing. *College Composition and Communication, 32,* 365-387.

Flower, L., Hayes, J. R., & Swarts, H. (1983). Revising functional documents: The scenario principle. In P. V. Anderson, R. J. Brockmann, & C. R. Miller (Eds.), *New essays in technical and scientific communication* (pp. 41-58). Farmingdale, NY: Baywood.

Flower, L., Wallace, D. L., Norris, L., & Burnett, R. E. (Eds.) (in press). *Making thinking visible: A collaborative look at collaborative planning.* Urbana, IL: NCTE.

Flynn, E. A. (1988). Composing as a woman. In S. L. Gabriel & I. Smithson (Eds.), *Gender in the classroom: Power and pedagogy* (pp. 112-126). Urbana, IL.: University of Illinois Press.

Flynn, E. A., Savage, G., Penti, M., Brown, C., & Watke, S. (1991). Gender and modes of collaboration in a chemical engineering design course. *Journal of Business and Technical Communication, 5,* 444-462.

Forman, J. (1990). Leadership dynamics of computer-supported writing groups. *Computers and Composition, 7,* 35-46.

Forman, J. (1991). Collaborative business writing: A Burkean perspective for future research. *Journal of Business Communication, 28,* 233-257.

Forman, J. (Ed.). (1992). *New visions of collaborative writing.* Portsmouth, NH: Boynton.

Forman, J., & Katsky, P. (1986). The group report: A problem in small group or writing processes? *Journal of Business Communication, 23*(4), 23-35.

Forty, A. (1986). *Objects of desire: Design and society 1750-1980.* New York: Pantheon Books.

Foss, S. K. (1989, November). *Implementing feminist pedagogy in the rhetorical criticism classroom.* Paper presented at the Speech Communication Association convention, San Francisco.

Foucault, M. (1971). *The archaeology of knowledge and the discourse on language* (A. M. Sheridan-Smith, Trans.). New York: Pantheon Books. (Original work published 1969)

Foucault, M. (1975). *Discipline and punish* (A. Sheridan, Trans.). New York: Pantheon Books.

Foucault, M. (1981). The order of discourse (I. McLeod, Trans.). In R. Young (Ed.), *Untying the text: A post-structuralist reader* (pp. 48-78). Boston: Routledge & Kegan Paul. (Original work published 1971)

Fox, T. (1990). *The social uses of writing: Politics and pedagogy.* Norwood, NJ: Ablex.

Freed, R. C., & Broadhead, G. J. (1987). Discourse conventions, sacred texts, and institutional norms. *College Composition and Communication, 38,* 154-165.

Freire, P. (1983). *The pedagogy of the oppressed* (M. B. Ramos, Trans.). New York: Continuum. (Original work published 1968)

Fresnault-Deruelle, P. (1983). *L'image manipulée.* Paris: Edilig.

Fulkerson, R. (1990). Composition theory in the eighties: Axiological consensus and paradigmatic diversity. *College Composition and Communication, 41,* 409-429.

Galegher, J., Kraut, R. E., & Egido, C. (Eds.). (1990). *Intellectual teamwork: Social and technological foundations of cooperative work.* Hillsdale, NJ: Lawrence Erlbaum.

Garland, K. (1969). The design of the London Underground Diagram. *The Penrose Annual, 62,* 68-82.

Garver, E. (1985). Teaching writing and teaching virtue. *Journal of Business Communication, 22*(1), 51-73.

Garver, E. (1987). *Machiavelli and the history of prudence.* Madison: University of Wisconsin Press.

Geertz, C. (1983). *Local knowledge: Further essays in interpretive anthropology.* New York: Basic Books.

Gere, A. R. (1987). *Writing groups: History, theory, and implications.* Carbondale: Southern Illinois University Press.

Gere, A. R., & Abbott, R. D. (1985). Talking about writing: The language of writing groups. *Research in the Teaching of English, 19,* 362-385.

Gerrety, J. (1991, January 25). Greeks OK stricter party rules. *Journal and Courier,* p. D1.

Gibson, J. J. (1954). A theory of pictorial perception. *Audio-Visual Communication Review, 1*, 3-23.

Giddens, A. (1979). *Central problems in social theory: Actions, structure and contradiction in social analysis.* Berkeley: University of California Press.

Giddens, A. (1987). *Social theory and modern sociology.* Stanford, CA: Stanford University Press.

Gilbert, G. N. (1977). Referencing as persuasion. *Social Studies of Science, 7,* 113-122.

Gilbert, G. N., & Mulkay, M. (1980). Contexts of scientific discourse: Social accounting in experimental papers. In K. D. Knorr, R. Krohn, & R. Whitley (Eds.), *The social process of scientific investigation* (pp. 269-294). Amsterdam: D. Reidel.

Gilbert, G. N, & Mulkay, M. M. (1984). *Opening Pandora's box: A sociological analysis of scientists' discourse.* Cambridge, UK: Cambridge University Press.

Gilligan, C. (1982). *In a different voice: Psychological theory and women's development.* Cambridge, MA: Harvard University Press.

Gilligan, C., Lyons, N. P., & Hanmer, T. J. (1990). Reflections: Conversations with Emma Willard teachers about their participation in the Dodge study. In C. Gilligan, N. P. Lyons, & T. J. Hanmer (Eds.), *Making connections: The relational worlds of adolescent girls at Emma Willard School* (pp. 286-313). Cambridge, MA: Harvard University Press.

Giroux, H. A. (1983). *Theory and resistance in education: A pedagogy for the opposition.* South Hadley, MA: Bergin & Garvey.

Gödel, K. (1962) *On formally undecidable propositions in "Principia mathematica" and related systems.* Edited by R. B. Braitewaite. New York: Basic Books.

Goetz, J. P., & LeCompte, M. D. (1984). *Ethnography and qualitative design in educational research.* San Diego: Academic Press.

Goldsmith, E. (1984). *Research into illustration: An approach and a review.* Cambridge, UK: Cambridge University Press.

Goldstein, J. [Rymer], & Malone, E. L. (1984). Journals on interpersonal and group communication: Facilitating technical projects groups. *Journal of Technical Writing and Communication, 14,* 113-131.

Goldstein, J. [Rymer], & Malone, E. L. (1985). Using journals to strengthen collaborative writing. *Bulletin of the Association for Business Communication, 48*(3), 24-28.

Golen, S., Powers, C., & Titkemeyer, M. A. (1985). How to teach ethics in a basic business communication class—committee report of the 1983 Teaching Methodology and Concepts Committee, Subcommittee 1. *Journal of Business Communication, 22*(1), 75-83.

Gombrich, E. H. (1972). The visual image. *Scientific American, 227*(3), 82-96.

Gombrich, E. H. (1979). *The sense of order: A study in the psychology of decorative art.* Ithaca, NY: Cornell University Press.

Goodman, N. (1976). *Languages of art: An approach to a theory of symbols* (2nd ed.). Indianapolis: Hackett.

Gorgias. (1990). *Helen* (G. A. Kennedy, Trans.). In P. P. Matsen, P. Rollinson, & M. Sousa (Eds.), *Readings from classical rhetoric* (pp. 34-36). Carbondale: Southern Illinois University Press. (Original work c.414 B.C.)

Gouran, D. S. (1986). Inferential errors, interaction, and group decision-making. In R. Y. Hirokawa & M. S. Poole (Eds.), *Communication and group decision-making* (pp. 93-111). Beverly Hills: Sage.

Gramsci, A. (1971). *Selections from the prison notebooks of Antonio Gramsci* (Q. Hoare & G. Smith, Eds. and Trans.). London: Lawrence and Wishart and International.

Green, B. (1987). Gender, genre, and writing pedagogy. In I. Reid (Ed.), *The place of genre in learning* (pp. 83-90, Typereader Publications no. 1). Geelong, Australia: Deakin University Press, The Centre for Studies in Literary Education.

Greene, S. (1990). Toward a dialectical theory of composing. *Rhetoric Review, 9,* 149-172.

Grice, R. A. (1991). Verifying technical information: Issues in information-development collaboration. In M. M. Lay & W. M. Karis (Eds.), *Collaborative writing in industry: Investigations in theory and practice* (pp. 224-241). Amityville, NY: Baywood.

Groshens, J. C. (Ed.). (1980). *Cartes et figures de la terre.* Paris: Centre Georges Pompidou.

Gross, A. G. (1990). *The rhetoric of science.* Cambridge, MA: Harvard University Press.

Group *Mu* (Eds.). (1978). *Collages.* Paris: Union Général d'Éditions.

Grumet, M. R. (1988). *Bitter milk: Women and teaching.* Amherst: University of Massachusetts Press.

Habermas, J. (1970). On systematically distorted communication. *Inquiry, 13,* 205-218.

Hacker, S. (1990). *Doing it the hard way: Investigations of gender and technology* (D. E. Smith & S. M. Turner, Eds.). Boston: Unwin Hyman.

Hackman, J. R. (1968). Effects of task characteristics on group products. *Journal of Experimental Social Psychology, 4,* 162-187.

Hackman, J. R., & Vidmar, N. (1970). Effects of size and task type on group performance and member reactions. *Sociometry, 33*(1), 37-54.

Hagan, M. A., & Jones, R. K. (1978). Cultural effects on pictorial perception: How many words is one picture really worth? In R. D. Walk & H. L. Pick, Jr. (Eds.), *Perception and experience* (pp. 171-212). New York: Plenum .

Hagge, J., & Kostlenick, C. (1989). Linguistic politeness in professional prose. A discourse analysis of auditors' suggestion letters, with implications for business communication pedagogy. *Written Communication, 6*(3), 312-339.

Hall, D. G., & Nelson, B. A. (1990). Sex-biased language and the technical-writing teacher's responsibility. *Journal of Business and Technical Communication, 4*(1), 69-79.

Hall, R. M., & Sandler, B. (1982). *The classroom climate: A chilly one for women?* (Project on the Status and Education of Women). Washington DC: Association of American Colleges.

Hall, S. (1982). The rediscovery of "ideology": Return of the repressed in media studies. In M. Gurevitch, T. Bennett, J. Curran, & J. Woolacott (Eds.), *Culture, society and the media* (pp. 56-90). London: Methuen.

Halliday, M. A. K. (1975). *Learning how to mean: Explorations in the development of language.* London: Edward Arnold.

Halliday, M. A. K. (1978). Poetry as scientific discourse: The nuclear sections of Tennyson's "In Memoriam." In D. Bioch & M. O'Toole (Eds.), *Functions of style* (pp. 31-44). London: F. Pinter.

Halliday, M. A. K., & Hasan, R. (1985). *Language, context and text: A social semiotic perspective.* Geelong, Australia: Deakin University Press.

Harcourt, J. (1990a). Developing ethical messages: A unit of instruction for the basic business communication course. *Bulletin of the Association for Business Communication, 53*(3), 17-20.

Harcourt, J. (1990b). Teaching the legal aspects of business communication. *Bulletin of the Association for Business Communication, 53*(3), 63-64.

Harkin, P. (1989). Bringing lore to light. *PRE/TEXT, 10*(1), 55-66.

Harley, J. B. (1988a). Maps, knowledge, and power. In D. Cosgrove & S. Daniels (Eds.), *The iconography of landscape: Essays on the symbolic representation, design and use of past environments* (pp. 277-312). Cambridge, UK: Cambridge University Press.

Harley, J. B. (1988b). Silences and secrecy: The hidden agenda of cartography in early modern Europe. *Imago Mundi, 40,* 57-76.

Harley, J. B. (1989). Historical geography and the cartographic illusion. *Journal of Historical Geography, 15*(1), 80-91.

Harris, J. (1989). The idea of community in the study of writing. *College Composition and Communication, 40,* 11-22.

Harris, R. C. (Ed.). (1987). *Historical atlas of Canada: Vol. 1. From the beginning to 1800.* Toronto: University of Toronto Press.

Harvey, D. (1989). *The condition of postmodernity: An enquiry into the origins of cultural change.* London: Basil Blackwell.

Hassan, I. (1987). Making sense: The trials of postmodern discourse. *New Literary History, 18,* 437-459.

Hayakawa, S. I. (1941.) *Language in thought and action.* New York: Harcourt, Brace & World.

Hayles, N. K. (1990). *Chaos bound: Orderly disorder in contemporary literature and science.* Ithaca, NY: Cornell University Press.

Hawisher, G. E., & Selfe, C. L. (1991). The rhetoric of technology and the electronic classroom. *College Composition and Communication, 42,* 55-65.

Heath, S. B. (1986). Critical factors in literacy development. In S. Castell et al. (Eds.), *Literacy society and schooling* (pp. 209-229). Cambridge, UK: Cambridge University Press.

Helgerson, R. (1986). The land speaks: Cartography, chorography, and subversion in Renaissance England. *Representations, 16,* 51-85.

Helgesen, S. (1990). *The female advantage: Women's ways of leadership.* New York: Doubleday.

Herrington, A. J. (1985). Classrooms as forums for reasoning and writing. *College Composition and Communication, 36*(4), 404-413.

Hillocks, G., Jr. (1986). *Research on written composition: New directions for teaching.* Urbana, IL: National Conference on Research in English.

Hobel, M. A. (1989). Ethics case study II: The boundaries of marketing integrity. *Intercom, 34*(7), 3, 9.

Huckin, T. N. (1987, March). *Surprise value in scientific discourse.* Paper presented at the annual meeting of the Conference on College Composition and Communication, Atlanta.

Hunter, J. (1990). Business ethics: Who cares? *Bulletin of the Association for Business Communication, 53*(3), 4-6.

Hutcheon, L. (1989). *The politics of postmodernism.* London: Routledge.

Hynds, S., & Rubin, D. L. (Eds.). (1990). *Perspectives on talk and learning.* Urbana, IL: National Council of Teachers of English.

Iser, W. (1978). *The act of reading: A theory of aesthetic response.* Baltimore: Johns Hopkins University Press.

Isocrates. (1929). Against the sophists. In Isocrates, *On the peace. Areopcagiticus. Against the sophists. Panathenaicus* (pp. 160-177; G. Norlin, Trans.). Cambridge, MA: Harvard University Press, Loeb Classical Library.

268   PROFESSIONAL COMMUNICATION

Jacobi, D., & Schiele, B. (1989). Scientific imagery and popularized imagery: Differences and similarities in the photographic portraits of scientists. *Social Studies of Science, 19,* 731-753.

Jameson, F. (1981). *The political unconscious: Narrative as a socially significant act.* Ithaca, NY: Cornell University Press.

Janis, I. L. (1972). *Victims of groupthink: A psychological study of foreign policy decisions and fiascoes.* Boston: Houghton Mifflin.

Janis, I. L. (1982). *Victims of groupthink* (2nd ed.). Boston: Houghton Mifflin.

Johannesen, R. L. (1990). Communication in organizations. In *Ethics in human communication* (3rd ed., pp. 151-168). Prospect Heights, IL: Waveland Press.

Johnson, D. W., & Johnson, R. T. (1979). Conflict in the classroom: Controversy and learning. *Review of Educational Research, 49,* 51-70.

Johnson, D. W. & Johnson, R. T. (1987). *Learning together and alone: Cooperative, competitive and individualistic learning* (2nd ed.). Englewood Cliffs, NJ: Prentice-Hall.

Jones, R. K., & Hagen, M. A. (1980). A perspective on cross-cultural picture perception. In M. A. Hagan (Ed.), *The perception of pictures: Vol. 2. Dürer's devices: Beyond the projective model of pictures* (pp. 193-226). New York: Academic Press.

Journal of Business Communication. (1990). [Special issue], *27*(3).

Journet, D. (1990). Forms of discourse and the sciences of the mind. Luria, Sacks, and the role of narrative in neurological case histories. *Written Communication, 7*(2), 171-199.

Kaplan, A., & Klein, R. (1985). *The relational self in late adolescent women* (Work in progress No. 17). Wellesley, MA: Wellesley College, The Stone Center for Developmental Services and Studies.

Kaiser, W. L. (1987). *A new view of the world.* New York: Friendship Press.

Karis, W. (1989). Conflict in collaboration: A Burkean perspective. *Rhetoric Review, 8*(1), 113-126.

Katz, J. (1988). The new geography: Diagrammatic cartography, orientation, and scale. *AIGA Journal of Graphic Design, 6*(2), 2.

Keller, E. F. (1985). *Reflections on gender and science.* New Haven, CT: Yale University Press.

Kellner, D. (1989). *Jean Baudrillard: From Marxism to postmodernism and beyond.* Cambridge, UK: Polity Press.

Kennedy, J. M. (1974). *A psychology of picture perception.* San Francisco: Jossey-Bass.

Kent, T. (1989a). Beyond system: The rhetoric of paralogy. *College English, 51,* 492-507.

Kent, T. (1989b). Paralogic hermeneutics and the possibilities of rhetoric. *Rhetoric Review, 8,* 24-42.

Kent, T. (1991). On the very idea of a discourse community. *College Composition and Communication, 42*(4), 425-446.

Kent, T. (in press-a). Externalism and the production of discourse. In T. Kent, *Paralogic rhetoric: Writing and reading as hermeneutic acts.* Lewisburg, PA: Bucknell University.

Kent, T. (in press-b). Interpretation and triangulation: A Davidsonian critique of reader-oriented literary theory. In R. W. Dasenbrock (Ed.), *Literary theory after Davidson.* University Park: Pennsyvania State University Press

Kent, T. (in press-c). Ethnography and objectivity. In T. Kent, *Paralogic rhetoric: Writing and reading as hermeneutic acts.* Lewisburg, PA: Bucknell University Press.

Kiesler, S., Siegel, J., & McGuire, T. (1988). Social psychological aspects of computer-mediated communication. In I. Grief (Ed.), *Computer-supported cooperative work* (pp. 657-682).

Killingsworth, M. J., & Steffens, D. (1989). Effectiveness in the environmental impact statement. *Written Communication, 6,* 155-180.

Kinneavy, J. L. (1986). *Kairos:* A neglected concept in classical rhetoric. In J. D. Moss (Ed.), *Rhetoric and praxis: The contribution of classical rhetoric to practical reasoning* (pp. 79-105). Washington, DC: Catholic University of America Press.

Knoblauch, C. H. (1988). Rhetorical constructions: Dialogue and commitment. *College English, 50*(2), 125-140.

Kohlberg, L. (1981). *The philosophy of moral development.* San Francisco: Harper & Row.

Konvitz, J. W. (1990). The nation-state, Paris, and cartography in eighteenth- and nineteenth-century France. *Journal of Historic Geography, 16*(1), 3-16.

Kostelnick, C. (1989). How readers perceive pictures: Generating design guidelines from empirical research. In *Proceedings of the 36th International Technical Communication Conference* (RT-47 to RT-50). Washington, DC: Society for Technical Communication.

Kramarae, C., & Treichler, P. A. (1990). Power relationships in the classroom. In S. L. Gabriel & I. Smithson (Eds.), *Gender in the classroom: Power and pedagogy* (pp. 41-59). Urbana: University of Illinois Press.

Kress, G. R. (1985). *Linguistic processes in sociocultural practice.* Geelong, Australia: Deakin University Press.

Kress, G. R. (1987). Genre in a social theory of language: A reply to John Dixon. In I. Reid (Ed.), *The place of genre in learning* (pp. 35-45, Typereader Publications no. 1). Geelong, Australia: Deakin University Press, The Centre for the Study of Literary Education.

Kuhn, T. (1962/1970). *The structure of scientific revolutions* (2nd ed.). Chicago: University of Chicago Press.

Kuspit, D. B. (1989). Collage: The organizing principle of art in the age of the relativity of art. In K. Hoffman (Ed.), *Collage: Critical views* (pp. 39-57). Ann Arbor, MI: UMI Research Press.

Lakatos, I. (1970). Falsification and the methodology of scientific research programs. In I. Lakatos & A. Musgrave (Eds.), *Criticism and the growth of knowledge* (pp. 91-196). Cambridge, UK: Cambridge University Press.

Langer, J. A. (1987). A sociocognitive perspective on literacy. In J. A. Langer (Ed.), *Language, literacy and culture: Issues of society and schooling* (pp. 1-20). Norwood, NJ: Ablex.

Langley, B. (1728). *New principles of gardening: Or, the laying out and planting, parterres, groves, wildernesses, labyrinths, avenues, parks, etc. . . .* London: Printed for A. Bettesworth and J. Batley.

Latour, B. (1987). *Science in action: How to follow scientists and engineers through society.* Cambridge, MA: Harvard University Press.

Latour, B., & Woolgar, S. (1986). *Laboratory life: The construction of scientific fl1111l acts.* Princeton, NJ: Princeton University Press.

Lauer, J., & Asher, J. W. (1988). *Composition research: Empirical designs.* New York: Oxford University Press.

Lay, M. M. (1989). Interpersonal conflict in collaborative writing: What we can learn from gender studies. *Journal of Business and Technical Communication, 3*(2), 5-28.

Lay, M. M., & Karis, W. M. (Eds.). (1991). *Collaborative writing in industry.* Amityville, NY: Baywood.

LeFevre, K. B. (1987). *Invention as a social act.* Carbondale: Southern Illinois University Press.

Lewis, P. V., & Speck, H. E., III. (1990). Ethical orientations for understanding business ethics. *Journal of Business Communication, 27,* 213-232.

Lewontin, R. C. (1991). Facts and factitious in natural sciences. *Critical Inquiry, 18,* 140-154.

Leydesdorff, L., & Amsterdamska, O. (1990). Dimensions of citation analysis. *Science Technology, and Human Values, 15,* 305-335.

Lincoln, Y. S., & Guba, E. G. (1985). *Naturalistic inquiry.* Beverly Hills: Sage.

Lipson, C. (1988). A social view of technical writing. *Journal of Business and Technical Communication, 2*(1), 7-20.

Locker, K. O. (1992). What makes a collaborative writing team successful? A case study of lawyers and social service workers in a state agency. In J. Forman (Ed.), *New visions of collaborative writing* (pp. 37-62). Portsmouth, NH: Boynton.

Louth, R. (1989). The tinkertoy project. In R. Louth & A. M. Scott (Eds.), *Collaborative technical writing: Theory and practice* (pp. 229-234). Minneapolis: Association for Teachers of Technical Writing.

Louth, R., & Scott, A. M. (Eds.). (1989). *Collaborative technical writing* (Association of Teachers of Technical Writing Anthology Series). Minneapolis: Association of Teachers of Technical Writing.

Lupton, E. (1986). Reading isotype. *Design Issues, 3*(2), 47-58.

Lynch, M., & Woolgar, S. (1988). Introduction: Sociological orientations to representational practice in science. *Human Studies, 11,* 99-116.

Lyons, N. P. (1990). Listening to the voices we have not heard: Emma Willard girls' ideas about self, relationships, and morality. In C. Gilligan, N. P. Lyons, & T. J. Hanmer (Eds.), *Making connections: The relational worlds of adolescent girls at Emma Willard School* (pp. 30-72). Cambridge, MA: Harvard University Press.

Lyons, N. P., Saltonstall, J. F., & Hanmer, T. J. (1990). Competencies and visions. In C. Gilligan, N. P. Lyons, & T. J. Hanmer (Eds), *Making connections: The relational worlds of adolescent girls at Emma Willard School* (pp. 183-214). Cambridge, MA: Harvard University Press.

Lyotard, J.-F. (1984). *The postmodern condition: A report on knowledge: Vol. 3. Theory and history of literature* (G. Bennington & B. Massumi, Trans.). Minneapolis: University of Minnesota Press.

Maher, F. (1985). Classroom pedagogy and the new scholarship on women. In M. Culley & C. Portuges (Eds.), *Gendered subjects: The dynamics of feminist teaching* (pp. 29-48). Boston: Routledge & Kegan Paul.

Mangan, J. (1978). Cultural conventions of pictorial representation: Iconic literacy and education. *Educational Communication and Technology: A Journal of Theory, Research, and Development, 26,* 245-267.

Marin, L. (1980). Les voies de la carte. In J.-C. Groshens (Ed.), *Cartes et figures de la terre* (pp. 47-54). Paris: Centre Georges Pompidou.

Marin, L. (1984). *Utopics: Spatial play* (R. A. Vollrath, Trans.). Atlantic Highlands, NJ: Humanities Press.

Marin, L. (1988). *Portrait of the king* (M. M. Houle, Trans.). Minneapolis: University of Minnesota Press.

Marius, R. (1990, Fall). On academic discourse. *Association of Departments of English Bulletin, 96,* 4-7.

Martin, J. R. (1986). Grammaticalizing ecology: The politics of baby seals and kangaroos. In T. Threadgold, E. A. Grosz, G. Kress, & M. A. K. Halliday (Eds.), *Semiotics, ideology, language* (pp. 225-267). Sidney, Australia: Sidney Association for Studies in Society and Culture.

Martin, J. R., Christie, F., & Rothery, J. (1987). Social processes in education: A reply to Sawyer and Watson (and others). In I. Reid (Ed.), *The place of genre in learning* (pp. 58-82, Typereader Publications no. 1). Geelong, Australia: Deakin University Press, The Centre for Studies in Literary Education.

Martin, J., Feldman, M., Hatch, M. J., & Sitkin, S. B. (1983). The uniqueness paradox in organizational stories. *Administrative Science Quarterly, 28,* 438-453.

Mathes, J. C., & Stevenson, D. W. (1991). *Designing technical reports: Writing for audiences in organizations* (2nd ed.). New York: Macmillan. (First edition published in 1976 by Bobbs-Merrill)

McCord, E. A. (1991). The business writer, the law, and routine business communication: A legal and rhetorical analysis. *Journal of Business and Technical Communication, 5,* 173-199.

McIntosh, P. (1985). *Feeling like a fraud* (Work in progress no. 18). Wellesley, MA: Wellesley College, The Stone Center for Developmental Services and Studies.

Medawar, P. B. (1964, August 1). Is the scientific paper fraudulent? *Saturday Review,* pp. 42-43.

Mendelson, M. (1989). The rhetorical case: Its Roman precedent and the current debate. *Journal of Technical Writing and Communication, 19,* 203-226.

Merton, R. K. (1968). *Social theory and social structure.* New York: Free Press.

Miller, C. & Swift, K. (1980). *Handbook of nonsexist writing: For writers, editors and speakers.* New York: Lippincott & Crowell.

Miller, C. R. (1979). A humanistic rationale for technical writing. *College English, 40,* 610-617.

Miller, C. R. (1984). Genre as social action. *Quarterly Journal of Speech, 70*(2), 151-167.

Miller, C. R. (1985). Invention in technical and scientific discourse. In M. G. Moran & D. Journet (Eds.), *Research in technical communication: A bibliographic sourcebook* (pp. 117-162). Westport, CT: Greenwood Press.

Miller, C. R. (1989). What's practical about technical writing? In B. E. Fearing & W. K. Sparrow (Eds.), *Technical writing: Theory and practice* (pp. 14-24). New York: Modern Language Association.

Miller, C. R., & Selzer, J. (1985). Special topics of argument in engineering reports. In L. Odell & D. Goswami (Eds.), *Writing in nonacademic settings* (pp. 309-341). New York: Guilford.

Miller, R. B., & Heiman, S. E. (1985). *Strategic selling.* New York: Warner Books.

Minerbrook, S. (1991, April 15). Mental maps: The politics of cartography. *U.S. News & World Report,* p. 60.

Mitchell, J. H., & Smith, M. K. (1989). The prescriptive versus the heuristic approach in teaching technical communication. In B. E. Fearing & W. K. Sparrow (Eds.), *Technical writing: Theory and practice* (pp. 115-127). New York: Modern Language Association.

Mitchell, W. J. T. (1987). Going too far with the sister arts. In J. A. W. Heffernan (Ed.), *Space, time, image, sign: Essays on literature and the visual arts* (pp. 1-11). New York: Peter Lang.

Moffett, J. (1968). Teaching the universe of discourse. Boston: Houghton Mifflin.

Moffett, J. (1981a). *Active voice: A writing program across the curriculum.* Montclair, NJ: Boynton.

Moffett, J. (1981b). *Coming on center: Essays on English education.* Montclair, NJ: Boynton.

272   PROFESSIONAL COMMUNICATION

Monmonier, M. S. (1982). Cartography, geographic information, and public policy. *Journal of Geography in Higher Education, 6*(2), 99-107.

Monmonier, M. S. (1991). *How to lie with maps.* Chicago: University of Chicago Press.

Morgan, M., Allen, N., Moore, T., Atkinson, D., & Snow, C. (1987). Collaborative writing in the classroom. *Bulletin of the Association for Business Communication, 50*(3), 20-26.

Morgan, M., & Murray, M. (1991). Insight and collaborative writing. In M. M. Lay & W. M. Karis (Eds.), *Collaborative writing in industry* (pp. 64-81). Amityville, NY: Baywood.

Mumby, D. K. (1988). *Communication and power in organizations: Discourse, ideology, and domination.* Norwood, NJ: Ablex.

Myers, G. (1985). The social construction of two biologists' proposals. *Written Communication, 2*(3), 219-245.

Myers, G. (1986). Reality, consensus, and reform in the rhetoric of composition teaching. *College English, 48,* 154-174.

Myers, G. (1990). *Writing biology: Texts in the social construction of scientific knowledge.* Madison: University of Wisconsin Press.

National Broadcasting Company. (1990, November 18). *NBC News Today* (7:00 a.m. EST).

Nelson, J. S., Megill, A. & McCloskey, D. N. (Eds.). (1987). *The rhetoric of the human sciences: Language and argument in scholarship and public affairs.* Madison: University of Wisconsin Press.

Neurath, O. (1980). *International picture language* (R. Kinross, Ed.; M. Neurath, Trans.). Reading, UK: University of Reading Department of Typography and Graphic Communication. (Original work published 1936)

Nietzsche, F. (1979). On truth and lies in a nonmoral sense. In D. Breazeale (Ed. and Trans.), *Philosophy and truth: Selections from Nietzsche's notebooks of the early 1870's* (pp. 79-100). Atlantic Highlands, NJ: Humanities Press.

Nietzsche, F. (1989). Ancient rhetoric. In S. L. Gilman, C. Blair, & D. J. Parent (Eds. and Trans.), *Friedrich Nietzsche on rhetoric and language* (pp. 1-193). New York: Oxford University Press.

Noddings, N. (1984). *Caring: A feminine approach to ethics & moral education.* Berkeley: University of California Press.

North, S. (1987). *The making of knowledge in composition: Portrait of an emerging field.* Upper Montclair, NJ: Boynton.

Novitz, D. (1977). *Pictures and their use in communication: A philosophical essay.* The Hague: Martinus Nijhoff.

Nystrand, M. (1986). *The structure of written communication: Studies in reciprocity between writers and readers.* Orlando, FL: Academic Press.

Nystrand, M. (1989). A social-interactive model of writing. *Written Communication, 6*(1), 66-85.

Nystrand, M. (1990). Sharing words: The effects of readers on developing writers. *Written Communication 7,* 3-24.

Odell, L. (1985). Beyond the text. Relations between writing and social context. In L. Odell & D. Goswami (Eds.), *Writing in nonacademic settings* (pp. 249-279). New York: Guilford.

Odell, L., & Goswami, D. (Eds.). (1985). *Writing in nonacademic settings.* New York: Guilford.

Ong, W. (1975). The writer's audience is always a fiction. *PMLA, 90,* 9-21.

Ormeling, F. (1980). Toponymies. In J.-C. Groshens (Ed.), *Cartes et figures de la terre* (pp. 332-334). Paris: Centre Georges Pompidou.

Palmer, S. E. (1975). Visual perception and world knowledge: Notes on a model of sensory-cognitive interaction. In D. A. Norman, D. E. Rumelhart, and the LNR Research Group. *Explorations in cognition* (pp. 279-307). San Francisco: Freeman.

Paradis, J., Dobrin, D., & Miller, R. (1985). Writing at Exxon ITD: Notes on the writing environment of an R&D organization. In L. Odell & D. Goswami (Eds.), *Writing in nonacademic settings* (pp. 281-307). New York: Guilford.

Park, D. B. (1986). Analyzing audiences. *College Composition and Communication, 37,* 478-488.

Parsons, G. M. (1987). Ethical factors influencing curriculum design and instruction in technical communication. *IEEE Transactions on Professional Communication, 30*(3), 202-207.

Pauley, S. E., & Riordan, D. G. (1990). *Technical report writing today* (4th ed.), Boston: Houghton Mifflin.

Perelman, C., & Olbrechts-Tyteca, L. (1958/1969). *The new rhetoric: A treatise on argumentation* (J. Wilkinson & P. Weaver, Trans.). Notre Dame, IN: University of Notre Dame Press.

Perkins, D. N. (1980). Pictures and the real thing. In P. A. Kolers, M. E. Wrolstad, & H. Bouma (Eds.), *Processing of visible language 2* (pp. 259-278). New York: Plenum.

Perry, W. (1968). *Forms of intellectual and ethical development in the college years.* New York: Holt, Rinehart & Winston.

Peters, A. (1983). *The new cartography.* New York: Friendship Press.

Peterson, J. (1991). Valuing teaching: Assumptions, problems and possibilities. *College Composition and Communication, 42,* 25-35.

Pettit, J. D., Jr., Vaught, B., & Pulley, K. J. (1990). The role of communication in organizations: Ethical considerations. *Journal of Business Communication, 27,* 233-249.

Phelps, L. W. (1988). *Composition as a human science: Contributions to the self-understanding of a discipline.* New York: Oxford University Press.

Plato. (1971). *Cratylus* (E. Hamilton & H. Cairns, Eds.). Princeton, NJ: Princeton University Press. (Original work c. 360 B.C.)

Plato. (1985a). *Gorgias* (W. C. Helmbold, Trans.). New York: Library of Liberal Arts. (Original work c.386 B.C.; trans. first published in 1953)

Plato. (1985b). *Phaedrus* (W. C. Helmbold & W. G. Rabinowitz, Trans.). New York: Library of Liberal Arts. (Original work c.370 B.C.; trans. first published in 1956)

Porter, J. E. (1987). Truth in technical advertising: A case study. *IEEE Transactions on Professional Communication, 30*(3), 182-189.

Porter, J. E. (1989, April). *Developing an ethical perspective in business writing.* Paper presented at Midwest Regional Conference, Association for Business Communication, Covington, KY.

Porter, J. E. (1990a). Ideology and collaboration in the classroom and in the corporation. *The Bulletin of the Association for Business Communication, 53*(2), 18-22.

Porter, J. E. (1990b, March). *The influence of the lawsuit on corporate composing processes.* Paper presented at Conference on College Composition and Communication, Chicago, IL.

Porter, J. E. (1990c, April). *Why teach public policy writing?* Paper presented at Midwest Regional Conference, Association for Business Communication, Detroit, MI.

Porter, J. E. (1991, March). *The roles of policy and critique in professional writing.* Paper presented at Conference on College Composition and Communication, Boston, MA.

Porter, J. E. (1992). *Audience and rhetoric: An archaeological composition of the discourse community.* Englewood Cliffs, NJ: Prentice Hall.

Pratt, M. L. (1977). *Towards a speech-act theory of literary discourse.* Bloomington, IN: Indiana University Press.

Pratt, S. (1990, November 1). Uniform food labeling rules should help consumers learn more about what they eat. *Chicago Tribune,* sec. 2, p. 7.

Prior, O. H. (1913). *Caxton's mirrour of the world* (Early English Text Society Extra Series Microfiche No. 110). Washington DC: NCR Microcard Editions.

Prodicus. (1972). Fragments. In R. K. Sprague (Ed. and Trans.), *The older sophists* (pp. 70-85). Columbia: University of South Carolina Press. (Original work c. 450 B.C.)

Protagoras. (1972). Fragments. In R. K. Sprague (Ed. and Trans.), *The older sophists* (pp. 3-28). Columbia: University of South Carolina Press. (Original work c.450 B.C.)

Purdue University. (1990). *Toward a university free of alcohol and drugs.* West Lafayette, IN: Office of Publications.

Putnam, L. L. (1986). Conflict in group decision-making. In R. Y. Hirokawa & M. S. Poole (Eds.), *Communication and group decision-making* (pp. 175-196). Beverly Hills: Sage.

Raffestin, C. (1980). *Pour une géographie du pouvoir.* Paris: Librairies Techniques.

Rafoth, B. A. (1988). Discourse community: Where writers, readers, and texts come together. In B. A. Rafoth & D. L. Ruben (Eds.), *The social construction of written communication* (pp. 131-146). Norwood, NJ: Ablex.

Rafoth, B. A. (1990). The concept of discourse community. In G. Kirsch & D. H. Roen (Eds.), *A sense of audience in written communication* (pp. 140-152). Newbury Park: Sage.

Raven, M. E. (1990). New venture techniques in a communication class. *Technical Writing Teacher, 17*(2), 124-130.

Reid, I. (Ed.). (1987). *The place of genre in learning* (Typereader Publications no. 1). Geelong, Australia: Deakin University Press, The Centre for Studies in Literary Education.

Reinsch, N. L., Jr. (1990). Ethics research in business communication: The state of the art. *Journal of Business Communication, 27,* 251-272.

Rentz, K. C., & Debs, M. B. (1987). Language and corporate values: Teaching ethics in business writing courses. *Journal of Business Communication, 24*(3), 37-48.

Riffaterre, M. (1978). *Semiotics of poetry.* Bloomington, IN: Indiana University Press.

Riipi, L., & Carlson, E. (1990). Tumor necrosis factor (TNF) is induced in mice by *Candida albicans:* Role of TNF in fibrinogen increase. *Infection and Immunity, 58,* 5019-5129.

Robinson, A. H. (1985). Arno Peters and his new cartography. *The American Cartographer, 12,* 103-111.

Rogers, P. S. (1989). Choice-based writing in managerial contexts: The case of the dealer contract. *The Journal of Business Communication, 26,* 197-216.

Rogers, P. S., & Horton, M. (1992). Face-to-face collaborative writing. In J. Forman (Ed.), *New visions of collaborative writing* (pp. 120-146). Portsmouth, NH: Boynton.

Rogers, P. S., & Swales, J. M. (1990). We the people? An analysis of the Dana Corporation policies document. *Journal of Business Communication, 27,* 293-313.

Rorty, R. (1979). *Philosophy and the mirror of nature.* Princeton, NJ: Princeton University Press.

Rorty, R. (1989). *Contingency, irony, and solidarity.* New York: Cambridge University Press.

Rothery, J. (1984). The development of genres—middle to junior secondary. In Deakin Board of Education (Ed.), *Children writing* (pp. 31-46, study guide). Geelong, Australia: Deakin University Press.

Sapir, E. (1921). *Language: An introduction to the study of speech*. New York: Harcourt, Brace & World.

Sawyer, W., & Watson, K. (1987). Questions of genre. In I. Reid (Ed.), *The place of genre in learning* (pp. 46-57, Typereader Publications no. 1). Geelong, Australia: Deakin University Press, The Centre for Studies in Literary Education.

Scanlon, R. (n.d.) Program notes to the American Reporting Theatre production of *The Writing Game*. Cambridge, MA.

Schall, M. S. (1983). A communication-rules approach to organizational culture. *Administrative Science Quarterly, 28*, 557-581.

Scharton, M. (1989). Models of competence: Responses to a scenario writing assignment. *Research in the Teaching of English, 23*, 163-180.

Scheibal, W. J. (1986). The effectiveness of plain English laws: A legal perspective. *Journal of Business Communication, 23*(3), 57-63.

Schneidewind, N. (1983). Feminist values: Guidelines for teaching methodology in women's studies. In C. Burch & S. Pallock (Eds.), *Learning our way: Essays in feminist education* (pp. 261-271). Trumansburg, NY: Crossing Press.

Scott, A. M. (1988). Group projects in technical writing courses. *The Technical Writing Teacher, 15*(2), 138-142.

Selfe, C. L., & Wahlstrom, B. J. (1989). Computer-supported writing classes: Lessons for teachers. In C. L. Selfe, D. Rodrigues, & W. R. Oates (Eds.), *Computers in English and the language arts* (pp. 257-268). Urbana, IL: National Council of Teachers of English.

Selzer, J. (1989). Composing processes for technical discourse. In B. E. Fearing & W. K. Sparrow (Eds.), *Technical writing: Theory and practice* (pp. 41-51). New York: Modern Language Association.

Shannon, C. E. (1948, July & August). A mathematical theory of information. *Bell System Technical Journal 27*, 379-423, 623-656.

Shapiro, M. J. (1988). *The politics of representation: Writing practices in biography, photography, and policy analysis*. Madison: University of Wisconsin Press.

Shaw, M. (1981). *Group dynamics*. New York: McGraw Hill.

Shirk, H. N. (Ed.). (1990). Ethics and Values [Special issue]. *Bulletin of the Association for Business Communication, 53*(3).

Simon, C. (1987, November 15). Place maps. *Psychology Today*, p. 15.

Simons, H. W. (Ed.). (1989). *Rhetoric in the human sciences*. Newbury Park, CA: Sage.

Simpson, M. (1991). Writing computer documentation for multiple audiences: An ethnographic study. Unpublished manuscript. West Lafayette, IN: Purdue.

Slavin, R. E. (1990). *Cooperative learning*. Englewood Cliffs, NJ: Prentice Hall.

Sless, D. (1986). Reading semiotics. *Information Design Journal, 4*(3), 179-189.

Small, H. G. (1977). A co-citational model of a scientific specialty: A longitudinal study of collagen research. *Social Studies of Science, 7*, 139-166.

Small, H. G. (1978). Cited documents as concept symbols. *Social Studies of Science, 8*, 327-340.

Smeltzer, L. R., & Werbel, J. D. (1986). Gender differences in managerial communication: Fact or folk-linguistics? *Journal of Business Communication, 23*(2), 41-50.

Smithson, I. (1990). Introduction: Investigating gender, power, and pedagogy. In S. L. Gabriel & I. Smithson (Eds.), *Gender in the classroom: Power and pedagogy* (pp. 1-27). Urbana: University of Illinois Press.

Souther, J. W. (1989). Teaching technical writing: A retrospective appraisal. In B. Fearing & W. K. Sparrow (Eds.), *Technical writing: Theory and practice* (pp. 2-13). New York: Modern Language Association.

Speck, B. W. (1990). Writing professional codes of ethics to introduce ethics in business writing. *Bulletin of the Association for Business Communication, 53*(3), 36-52.

Speck, B. W., & Porter, L. R. (1990). Annotated bibliography for teaching ethics in professional writing. *Bulletin of the Association for Business Communication, 53*(3), 36-51.

Spencer, B. A., & Lehman, C. M. (1990). Analyzing ethical issues: Essential ingredient in the business communication course of the 1990s. *Bulletin of the Association for Business Communication, 53*(3), 7-16.

Spilka, R. (1988). Studying writer-reader interactions in the workplace. *Technical Writing Teacher, 15*(3), 208-221.

Spilka, R. (1990). Orality and literacy in the workplace: Process- and text-based strategies for multiple-audience adaptation. *Journal of Business and Technical Communication, 4,* 44-67.

Steiner, I. D. (1972). *Group process and productivity.* New York: Academic Press.

Sterkel, K. S. (1988). The relationship between gender and writing style in business communication. *Journal of Business Communication, 25*(4), 17-38.

Stern, L. (1990). Conceptions of separation and connection in female adolescents. In C. Gilligan, N. P. Lyons, & T. J. Hanmer (Eds.), *Making connections: The relational worlds of adolescent girls at Emma Willard School* (pp. 73-87). Cambridge, MA: Harvard University Press.

Stevenson, D. W. (1986, October). *The legal context: Issues for technical communicators.* Paper presented at the roundtable for teachers of business and technical writing, Purdue University, West Lafayette, IN.

Suchman, L. A. (1987). *Plans and situated actions: The problem of human-machine communication.* Cambridge, UK: Cambridge University Press.

Sullivan, D. L. (1990). Political-ethical implications of defining technical communication as a practice. *Journal of Advanced Composition, 10,* 375-386.

Sullivan, P., & Porter, J. E. (in press). On theory, practice, method: Toward a heuristic research methodology. In R. Spilka (Ed.), *Writing in the workplace: New research perspectives.* Carbondale: Southern Illinois University Press.

Swales, J. (1986). Citation analysis and discourse analysis. *Applied Linguistics, 7,* 39-56.

Swales, J. (1990). *Genre analysis: English in academic and research settings.* Cambridge, UK: Cambridge University Press.

Szwed, J. F. (1981). The ethnography of literacy. In M. F. Whiteman (Ed.), *Writing: The nature, development, and teaching of written communication* (pp. 13-23). Hillsdale, NJ: Lawrence Erlbaum.

Taber, H. G., Naeve, L., Agnew, M., & Heer, R. (1991). *Composting yard waste* (rev. ed., Pm-683). Ames: Iowa State University Extension.

Tackach, J. (1991). Teaching students to manage documents. *Technical Communication, 38*(1), 118-119.

Tannen, D. (1990). *You just don't understand: Women and men in conversation.* New York: William Morrow.

Tebeaux, E. (1980). Let's not ruin technical writing, too: A comment on the essays of Carolyn Miller and Elizabeth Harris. *College English, 41,* 822-825.

Tebeaux, E. (1988). The trouble with employees' writing may be freshman composition. *Teaching English in the Two-Year College, 15*(1), 9-19.

Tebeaux, E. (1990a). *Design of business communications: The process and the product.* New York: Macmillan.

Tebeaux, E. (1990b). Toward an understanding of gender differences in written business communications: A suggested perspective for future research. *Journal of Business and Technical Communication, 4*(1), 25-43.

Tebeaux, E. (1991). The shared-document collaborative case response. In M. M. Lay & W. M. Karis (Eds.), *Collaborative writing in industry* (pp. 124-145). Amityville, NY: Baywood.

Thompson, I. (1991). The speech community in technical communication. *Journal of Technical Writing and Communication, 21*(1), 41-54.

Thompson, J. B. (1984). *Studies in the theory of ideology.* Cambridge, UK: Polity Press.

Threadgold, T., Grosz, E. A., Kress, G. & Halliday, M. A. K. (Eds.). (1986). *Semiotics, ideology, language.* Sydney, Australia: Sydney Association for Studies in Society and Culture.

Tirrell, M. K., Pradl, G. M., Warnock, J, & Britton, J. (1990). Representing James Britton: A symposium. *College Composition and Communication, 41*(2), 166-186.

Toffler, A. (1990). *Powershift: Knowledge, wealth, and violence at the edge of the 21st century.* New York: Bantam.

Torgovnick, M. (1990, Fall). Experimental critical writing. *Association of Departments of English Bulletin, 96,* 8-10.

Toulmin, S. E. (1983). *The uses of argument.* Cambridge, UK: Cambridge University Press.

Trecker, J. L. (1974). Sex, science and education. *American Quarterly, 26*(4), 352-366.

Treichler, P. A., & Frank, F. W. (1989). Guidelines for nonsexist usage. In F. W. Frank & P. A. Treichler (Eds.), *Language, gender, and professional writing: Theoretical approaches and guidelines for nonsexist usage* (pp. 137-226). New York: Modern Language Association.

Trice, H. M., & Beyer, J. M. (1984). Studying organizational cultures through rites and ceremonials. *Academy of Management Review, 9,* 653-669.

Trieb, M. (1980). Mapping experience. *Design Quarterly, 115,* 5-32.

Trimbur, J. (1989). Consensus and difference in collaborative learning. *College English, 51,* 602-616.

Trimbur, J., & Braun, L. A. (1992). Laboratory life and the determination of authorship. In J. Forman (Ed.), *New visions of collaborative writing* (pp. 19-36). Portsmouth, NH: Boynton.

Tufte, E. R. (1983). *The visual display of quantitative information.* Cheshire, CT: Graphics Press.

Tufte, E. R. (1990). *Envisioning information.* Cheshire, CT: Graphics Press.

Twain, M. (1924). *Tom Sawyer abroad, Tom Sawyer, detective and other stories, etc., etc.* New York: Harper.

Twyman, M. (1985). Using pictorial language: A discussion of the dimensions of the problem. In T. M. Duffy & R. Waller (Eds.), *Designing usable texts* (pp. 245-312). New York: Academic.

Ulmer, G. L. (1983). The object of post-criticism. In H. Foster (Ed.), *The anti-aesthetic: Essays on postmodern culture* (pp. 83-110). Port Townsend, WA: Bay Press.

Uspensky, B. (1973). *A poetics of composition: The structure of the artistic text and typology of a compositional form* (V. Zavarin & S. Wittig, Trans.). Berkeley: University of California Press.

Van Maanen, J. (1988). *Tales of the field: On writing ethnography.* Chicago: University of Chicago Press.

Van Pelt, W., & Gillam, A. (1991). Peer collaboration and the computer-assisted classroom: Bridging the gap between academia and the workplace. In M. M. Lay & W. K. Karis (Eds.), *Collaborative writing in industry: Investigations in theory and practice* (pp. 170-205). Amityville, NY: Baywood.

Venturi, R. (1966). *Complexity and contradiction in architecture.* New York: Museum of Modern Art.

Vico, G. (1709/1965). *On the study methods of our time* (E. Gianturco, Ed. and Trans.). Indianapolis: Bobbs-Merrill.

Vico, G. (1744/1948). *The new science of Giambattista Vico* (T. Bergin & M. Fisch, Eds. and Trans.). Ithaca, NY: Cornell University Press.

Vygotsky, L. (1962). *Thought and language* (E. Hanfmann & G. Vakar, Trans.). Cambridge: MIT Press.

Vygotsky, L. (1986). *Thought and language* (A. Kozulin, Trans.). Cambridge: MIT Press.

Wahr, D. (1987, November 7). Personal communication with B. F. Barton & M. S. Barton.

Wakefield, N. (1990). *Postmodernism: The twilight of the real.* London: Pluto Press.

Walker, J. (1990). Of brains and rhetorics. *College English, 52,* 301-322.

Walker, J. A. (1979). The London Underground Diagram. *Icographic, 1*(14/15), 2-4.

Wall, V. D., & Nolan, L. L. (1987). Small group conflict: A look at equity, satisfaction, and styles of conflict management. *Small Group Behavior, 18,* 188-211.

Walzer, A. E. (1985). Articles from the "California divorce project": A case study of the concept of audience. *College Composition and Communication, 36,* 150-159.

Walzer, A. E. (1989). The ethics of false *implicature* in technical and professional writing courses. *Journal of Technical Writing and Communication, 19,* 149-160.

Weiler, K. (1988). *Women teaching for change: Gender, class, power* (Critical Studies in Education Series). South Hadley, MA: Bergin & Garvey.

Wells, S. (1986). Jürgen Habermas, communicative competence, and the teaching to technical discourse. In C. Nelson (Ed.), *Theory in the classroom* (pp. 245-269). Urbana: University of Illinois Press.

Whitehead, A. N., & Russell, B. A. W. (1910-1913). *Principia mathematica,* 3 vols. Cambridge: Cambridge University Press.

Whorf, B. (1956). *Language, thought, and reality.* New York: John Wiley.

Wiener, H. S. (1986). Collaborative learning in the classroom: A guide to evaluation. *College English, 48,* 52-61.

Williams, R. (1960). *Culture and society 1780-1950.* New York: Columbia University Press.

Williams, R. (1973). Base and superstructure in Marxist cultural theory. *New Left Review, 82,* 3-16.

Williamson, J. (1978). *Decoding advertisements: Ideology and meaning in advertisements.* London and New York: Marion Boyars.

Winograd, T., & Flores, C. F. (1986). *Understanding computers and cognition: A new foundation for design.* Norwood, NJ: Ablex.

Winsor, D. A. (1989). An engineer's writing and the corporate construction of knowledge. *Written Communication, 6*(3), 270-285.

Winsor, D. A. (1990a). The construction of knowledge in organizations: Asking the right questions about the Challenger. *Journal of Business and Technical Communication, 4*(2), 7-20.

References   279

Winsor, D. A. (1990b). Engineering writing: Writing engineering. *College Composition and Communication, 41,* 58-70.

Wood, D. (1987). Pleasure in the idea: The atlas as narrative form. *Cartographica, 24*(1), 24-45.

Wood, D., & Fels, J. (1986). Designs on signs: Myth and meaning in maps. *Cartographica, 23*(3), 54-103.

Worthen, W. E. (Ed.). (1869). *Appletons' cyclopaedia of drawing, designed as a text-book for the mechanic, architect, engineer, and surveyor . . .* New York: Appleton .

Yoos, G. E. (1984). Rational appeal and the ethics of advocacy. In R. J. Connors, L. S. Ede, & A. Lunsford (Eds.), *Essays on classical rhetoric and modern discourse* (pp. 82-97). Carbondale: Southern Illinois University Press.

Young, R. E., Becker, A. L., & Pike, K. L. (1970). *Rhetoric: Discovery and change.* New York: Harcourt.

Zappen, J. P. (1987). Rhetoric and technical communication: An argument for historical and political pluralism. *Journal of Business and Technical Communication, 1*(2), 29-44.

Zappen, J. P. (1991). Scientific rhetoric in the nineteenth and early twentieth centuries: Herbert Spencer, Thomas H. Huxley, and John Dewey. In C. Bazerman & J. Paradis (Eds.), *Textual dynamics of the professions: Historical and contemporary studies of writing in professional communities* (pp. 145-167). Madison: University of Wisconsin Press.

Zuckerman, S. H., & Bendele, A. M. (1989). Regulation of serum tumor necrosis factor in glucocorticoid-sensitive and -resistant rodent endotoxin shock models. *Infection and Immunity, 57,* 3009-3013.

Index

About the Contributors

Ben F. Barton is Professor of Electrical and Computer Engineering in the College of Engineering at the University of Michigan. **Marthalee S. Barton** is in the Program in Technical Communication at the University of Michigan. They were recipients of the 1988 NCTE Award for Excellence in Technical and Scientific Communication for their article "Simplicity in Visual Representation: A Semiotic Approach," which appeared in the *Journal of Business and Technical Communication.* They were also guest editors of the 1990 special issue of *The Technical Writing Teacher,* which was devoted to visual representation in technical and professional discourse, and won the 1991 NCTE award for Excellence in Technical and Scientific Communication as the "Best Collection of Essays." They are currently serving as guest editors of an issue of the *Journal of Business and Technical Communication* titled "Power and Professional Communication."

Carol Berkenkotter is Associate Professor of Rhetoric and Composition at Michigan Technological University, where she teaches courses on qualitative research methods in composition, composition theory, and writing for publication in the Doctoral Program in Rhetoric and Technical Communication. She has published a number of articles on the sociocognitive contexts of written communication and is co-authoring a book on genre knowledge in academic writing.

Nancy Roundy Blyler is Associate Professor at Iowa State University, where she teaches graduate and undergraduate courses in professional communication. She co-founded and co-edited the *Journal of Business and Technical Communication.* She has published in such journals as the *Journal of Technical Writing and Communication,* the *Journal of Business Communication, The Technical Writing Teacher,* and the *Journal of Advanced Composition.* An article she co-authored—"Real Readers, Implied Readers, and Professional Writers: Suggested Research"—won the 1989 Alpha Kappa Psi Foundation Award for Distinguished Publication on Business Communication.

Rebecca E. Burnett is Assistant Professor in the Department of English at Iowa State University, where she teaches technical communication and rhetorical analysis. She is also an Affiliated Researcher at the Center for the Study of Writing and Literacy, at the University of California at Berkeley and Carnegie Mellon University. Her primary research involves examining individual and social factors that affect co-authors' collaborative planning of documents. She has conducted numerous workshops and seminars for groups in business and industry and also serves as a consultant for national education projects. She is author or co-author of four texts in literature, composition, and technical communication, including *Technical Communication,* and several articles and book chapters.

Joseph J. Comprone is Head of the Department of Humanities at Michigan Technological University, where he is helping to develop doctoral and masters programs in rhetoric and technical communication. Previously, he was Director of Graduate Studies and Composition at the University of Louisville (where he developed and helped establish the doctoral program in rhetoric and composition), the University of Cincinnati, and the University of Minnesota at Morris. He spent a year (1988-1989) teaching and consulting with the linguistics and writing faculty at the National University of Singapore, where he contributed to the development of the university's new major concentration in composition and rhetoric. He has published more than 40 essays, chapters, reviews, and notes in professional journals, on subjects that include literary theory and writing, film and composition, and theories of literacy. He has also published four college writing textbooks and edited an anthology on composition teaching. Most recently, he has been working on a book on Kenneth Burke and composition studies and a collection of essays on writers such as Oliver Sacks who combine literary and scientific genres in their work.

Richard C. Freed, Associate Professor in the Department of English at Iowa State University, is co-author of *The Variables of Composition,* which won the 1987 NCTE Best Book Award for Excellence in Scientific and Technical Communication. He has published articles on rhetoric and professional communication in *College Composition and Communication,* the *Journal of Technical Writing and Communication,* and *The Technical Writing Teacher.* He is completing a book on the writing of consulting proposals and has conducted workshops on proposal writing throughout the United States, as well as in Canada, Japan, Singapore, Scandinavia, and India.

Bruce Herzberg is Associate Professor of English at Bentley College, where he directs the freshman writing and the writing across the curriculum programs and teaches composition, speech, rhetoric, and literature. He has published a number of articles on composition and rhetoric and is co-author of *The Rhetorical Tradition* and *The Bedford Bibliography for Teachers of Writing.* Herzberg is presently at work on a co-authored composition textbook called *Negotiating Difference.*

Thomas N. Huckin is Director of the Writing Program at the University of Utah, where he is also Associate Professor in the Department of English. He teaches courses in discourse analysis, stylistics, and technical and professional writing. He has written many articles on these topics and has recently co-authored a new textbook, *Technical Writing and Professional Communication.*

Thomas Kent is Associate Professor in the Department of English at Iowa State University. He is currently editor of the *Journal of Business and Technical Communication.* His research interests include rhetorical and literary theory, and he has published on both topics in a variety of journals such as *College English, Rhetoric Review,* and *Philosophy and Rhetoric.* His first book, *Interpretation and Genre: The Role of Generic Perception in the Study of Narrative Texts,* appeared in 1986, and his second, on paralogic rhetoric, is scheduled to appear in 1992.

Charles Kostelnick is Associate Professor of English at Iowa State University, where he has taught business and technical communication as well as a graduate course in visual communication in professional writing. He has published in the *Journal of Business and Technical Communication, The Technical Writing Teacher,* and *College Composition and*

Communication and has presented papers at several national conventions. Much of his work has focused on issues related to visual communication and design.

Janice M. Lauer is Professor of English at Purdue University, where she is Director of the Graduate Program in Rhetoric and Composition. She is co-author of *Four Worlds of Writing* and *Composition Research: Empirical Designs* and author of articles on writing as inquiry, composition as a discipline, and invention. For 13 years, she was Director of Current Theories in Teaching Composition, a summer rhetoric seminar. She has been Chair of the College Section of NCTE and a member of the Executive Committee of CCCC and MLA's discussion group on the history and theory of rhetoric. Her current research interests include historical studies of invention, persuasive writing as inquiry and critique, and pedagogies of invention.

Mary M. Lay is Associate Professor in the Department of Rhetoric at the University of Minnesota-Twin Cities. She is a former president of the Association of Teachers of Technical Writing. She is co-editor of the official ATTW journal, *Technical Communication Quarterly,* and co-editor of a 1991 collection of essays, *Collaborative Writing in Industry.* She won the 1990 NCTE Award for Best Article on Philosophy or Theory of Technical Writing for her essay "Interpersonal Conflict in Collaborative Writing: What We Can Learn from Gender Studies," which appeared in the *Journal of Business and Technical Communication,* and is the former Chair of the Department of Technical Communications at Clarkson University.

Meg Morgan, Assistant Professor at the University of North Carolina at Charlotte, teaches technical communication, composition, and graduate courses in rhetoric and composition theory and history. She has written articles on collaboration and has published in the *Journal of Business and Technical Communication, Technical Communication,* and the *Bulletin of the Association for Business Communication.* Her articles have also appeared in *Effective Documentation: What We Learn from Research* and *Collaborative Writing in Industry.* She recently co-edited a special issue of *Technical Communication* devoted to collaboration and is designing, with colleagues, a program in technical communication at UNC-Charlotte.

James E. Porter is Associate Professor of English and Director of Business Writing at Purdue University, where he teaches in the Graduate Rhetoric

Program as well as in the Undergraduate Professional Writing Major. His research bridges his interests in rhetoric theory (particularly of audience and inventional methodology) and professional writing practice. His book *Audience and Rhetoric: An Archaeological Composition of the Discourse Community* examines rhetorical theories of audience in light of postmodern theory. He has published in journals such as *Rhetoric Review, Technical Communication, IEEE Transactions on Professional Communication,* and the *Journal of Teaching Writing,* and he has chapters forthcoming in two collections.

Jone Rymer (formerly Goldstein) is Associate Professor in the School of Business Administration at Wayne State University. Recently, she was a visiting associate professor in the Anderson Graduate School of Management at UCLA. She has co-authored several articles on collaborative writing, co-edited a special issue of the *Bulletin of the Association for Business Communication* on collaboration, and co-authored *Cases for Technical and Professional Writing.* Her recent publications are on professional writing processes, writing assessment, and pedagogical methods.

Patricia Sullivan is Associate Professor of English at Purdue University, where she is Director of the Technical Writing Program. A winner of an NCTE Award for Best Article on Methods of Teaching Technical or Scientific Communication and first finalist for the Nold Award for Best Article on Computers and Writing, she has authored articles on technical and professional communication theory, computers and writing, the usability of documentation, qualitative research in the workplace, and electronic publishing in journals such as *IEEE Transactions on Professional Communication, The Technical Writing Teacher, College Composition and Communication, Technical Communication,* and the *Journal of Technical Writing and Communication.* She is the current Chair of the NCTE Committee on Technical and Scientific Communication, has revised the professional writing major at Purdue, and directed efforts in instructional computing. Her research interests are professional writing as a discipline, qualitative research in the workplace, computers and writing, and the role of theory in workplace literacy.

Charlotte Thralls is Associate Professor at Iowa State University where she teaches in both the Graduate and the Undergraduate Rhetoric and Professional Communication Programs. She was co-founder and co-editor of the *Journal of Business and Technical Communication,* and she has been co-recipient of the Alpha Kappa Psi Foundation Award for Distinguished Publication on Business Communication. Her current research explores social-based rhetorical theory in relation to organizational culture and visual media.